Mobile Technologies of the City

Mobile communications technologies are joining new systems of urban transportation, surveillance, scheduling and sorting to quietly but dramatically change
the social, architectural and infrastructural fabric of cities across the world.
Urbanism has always been in flux, but now more so than ever, as many aspects of
economic and social life are increasingly conducted 'on the move' or away from
'home'. Emergent forms of physical and informational mobility are changing and
influencing patterns of movement, co-presence, social exclusion and security across
many urban contexts.

In *Mobile Technologies of the City* Mimi Sheller and John Urry bring together
a carefully selected group of innovative case studies that trace the emergence both
of the new socio-technical practices of the city and of a new theoretical paradigm
for mobilities research. The research based in Vienna, Liverpool, Bristol, London,
Tokyo, Paris, Los Angeles and Hong Kong ranges across media including guidebooks, web sites, train schedules, wireless computing, cinema, mobile phones
and mobile gaming. It encompasses infrastructures such as road and rail systems,
airline networks and hubs, internet routers and wireless 'hot spots', and focuses on
urban sites including streets, train platforms, bus stops, airports, internet cafés, park
benches, parking garages and cars.

Mobile Technologies of the City contributes new theoretical perspectives on
temporality, security and systematization. It brings together complexity theory, actor
network theory and social studies of technology generating a new approach to the
mobility systems that constitute contemporary cities.

Mimi Sheller is Visiting Associate Professor at Swarthmore College in Pennsylvania and Senior Research Fellow at the Centre for Mobilities Research,
Lancaster University, England. She is the author of *Democracy After Slavery* (2000)
and *Consuming the Caribbean* (2003), and co-editor of *Uprootings/Regroundings*
(2003) and *Tourism Mobilitites* (2004). She is also co-editor of the new Routledge
journal *Mobilities*.

John Urry is Professor of Sociology and Director at the Centre for Mobilities
Research, Lancaster University. Recent books include *Sociology beyond Societies*
(2000), *The Tourist Gaze* (1990/2002), *Global Complexity* (2003), *Tourism Mobilities*
(co-edited, 2004) and *Automobilities* (co-edited, 2005). He is also co-editor of the
new Routledge journal *Mobilities*.

The Networked Cities Series

Series Editors:
Richard E. Hanley
New York City College of Technology, City University of New York, US
Steve Graham
Department of Geography, Durham University, UK
Simon Marvin
SURF, Salford University, UK

From the earliest time, people settling in cities devised clever ways of moving things: the materials they needed to build shelters, the water and food they needed to survive, the tools they needed for their work, the armaments they needed for their protection – and ultimately, themselves. Twenty-first century urbanites are still moving things about, but now they employ networks to facilitate that movement – and the things they now move include electricity, capital, sounds and images.

The Networked Cities Series has as its focus these local, global, physical, and virtual urban networks of movement. It is designed to offer scholars, practitioners and decision-makers studies on the ways cities, technologies and multiple forms of urban movement intersect and create the contemporary urban environment.

The Network Society
A New Context for Planning
Edited by Louis Albrechts and Seymour Mandelbaum

Moving People, Goods and Information in the 21st Century
The Cutting-Edge Infrastructures of Networked Cities
Edited by Richard E. Hanley

Digital Infrastructures
Enabling Civil and Environmental Systems through Information Technology
Edited by Rae Zimmerman and Thomas Horan

Sustaining Urban Networks
The Social Diffusion of Large Technical Systems
Edited by Olivier Coutard, Richard E. Hanley and Rae Zimmerman

Mobile Technologies of the City

Edited by
Mimi Sheller and John Urry

Routledge
Taylor & Francis Group

LONDON AND NEW YORK

First published 2006 by Routledge
2 Park Square, Milton Park, Abingdon, Oxon OX14 4RN

Simultaneously published in the USA and Canada
by Routledge
711 Third Avenue, New York, NY 10017

First issued in paperback 2012
Routledge is an imprint of the Taylor & Francis Group

© 2006 Edited by Mimi Sheller and John Urry

Typeset in Times New Roman by
Florence Production Ltd, Stoodleigh, Devon

British Library Cataloguing in Publication Data
A catalogue record for this book is available from the British Library

Library of Congress Cataloging in Publication Data
  Mobile technologies of the city/(edited by) Mimi Sheller and John Urry.
      p. cm. – (The networked cities series)
    1. City and town life. 2. City dwellers – Effect of technologies
  innovations on.   3. Social mobility.   4. Information technology –
  Social aspects.   5. Mobile communication systems – Social aspects.
  6. Urban transportation – Social aspects.   7. Movement (Philosophy).
  I. Sheller, Mimi.   II. Urry, John.   III. Series.
  HT108.M63 2006
  307.76 – dc22

ISBN13: 978-0-415-65560-6 (PBK)

ISBN13: 978-0-415-37434-7 (HBK)

CONTENTS

NOTES ON CONTRIBUTORS   vii

CHAPTER ONE   Introduction: Mobile Cities, Urban Mobilities   1
Mimi Sheller and John Urry

PART I
Mobilities and the Creation of Urban Spatial Form   19

CHAPTER TWO   The Linear City: Touring Vienna in the
Nineteenth Century   21
Ulrike Spring

CHAPTER THREE   Between the Physical and the Virtual:
Connected Mobilities?   44
Peter Adey and Paul Bevan

CHAPTER FOUR   Urban Violence: Luxury in Made Space   61
Sarah S. Jain

PART II
Re-configuring Co-presence   77

CHAPTER FIVE   Bypassing and WAPing: Reconfiguring
Timetables for 'Real-time' Mobility   79
Juliet Jain

CHAPTER SIX   Reshaping Patterns of Mobility and Exclusion?
The Impact of Virtual Mobility upon
Accessibility, Mobility and Social Exclusion   102
Susan Kenyon

CHAPTER SEVEN   Twin Towers and Amoy Gardens: Mobilities,
                  Risks and Choices   121
                  Stephen Little

PART III
Cultures of Infrastructure and Public Space   135

CHAPTER EIGHT   From Café to Park Bench: Wi-Fi® and
                  Technological Overflows in the City   137
                  Adrian Mackenzie

CHAPTER NINE   ICTs and the Engineering of Encounters:
                  A Case Study of the Development of a
                  Mobile Game Based on the Geolocation
                  of Terminals   152
                  Christian Licoppe and Romain Guillot

CHAPTER TEN   Permeable Boundaries in the Software-
                  sorted Society: Surveillance and
                  Differentiations of Mobility   177
                  *David Murakami Wood and
                  Stephen Graham*

INDEX   192

NOTES ON CONTRIBUTORS

Mimi Sheller is Visiting Associate Professor in Sociology and Anthropology, Swarthmore College, USA, and Visiting Senior Fellow in the Centre for Mobilities Research Lancaster University, England. She is the author of *Democracy After Slavery* (2000), *Consuming the Caribbean* (2003) and co-editor of *Uprootings/ Regroundings* (2003), *Tourism Mobilities* (2004), and of the new journal *Mobilities*. m.sheller@lancaster.ac.uk

John Urry is Professor of Sociology and Director of the Centre for Mobilities Research, Lancaster University. His recent books include *Sociology beyond Societies* (2000), *The Tourist Gaze* (1990/2002), *Global Complexity* (2003), *Tourism Mobilities* (co-edited; 2004), *Automobilities* (co-edited; 2005). He is the co-editor of the new journal *Mobilities*. j.urry@lancaster.ac.uk

Peter Adey is currently completing his Ph.D. 'Spaces of flow: mobility and identity in airport space' at the University of Wales, Aberystwyth. His research focuses upon mobilities, materiality and wider social theory. He has also published articles concerning airport control, surveillance and identity. pna98@aber.ac.uk

Paul Bevan is currently completing his Ph.D. at the University of Wales, Aberystwyth. His research includes the impacts of the continuing growth of ICT use, including those on the relationship between globalization and time and space, the cultural impacts of International ICTs, legality and boundaries in the age of the Internet, mobility and communication, digital divides, virtual interaction and post-humanism. ppb98@aber.ac.uk

Stephen Graham is Professor of Human Geography at Durham University. His research develops critical and 'socio-technical' perspectives to the reconfiguration of cities, technologies, mobility systems, and the relations between cities and organized, political violence. He is the co-author of *Telecommunications and the City*, *Splintering Urbanism*, co-editor of *Managing Cities*, and editor of *Cybercities Reader* and *Cities, War and Terrorism*. s.d.n.graham@durham.ac.uk

Romain Guillot is currently pursuing post-graduate studies in Paris concerning social studies of the varying use of mobile technologies. He is also involved in composing music and launching a start-up company using musical ringtones for promoting independent artists.

Juliet Jain completed her Ph.D. 'Networks of the future: time, space and rail travel' at Lancaster University before moving to the Centre for Transport and Society at the University of the West of England in 2003. Her research interests are oriented around the use of social theories of time and space, and actor-network theory, within the context of everyday mobility, social practice and transport policy. Juliet.Jain@uwe.ac.uk

Sarah S. Jain is Assistant Professor in Cultural and Social Anthropology at Stanford University, USA. Her research interests are broadly situated under the overlapping headings of design, injury and mobility. Her most recent book, *Injury*, analyses injury production in the US through the lens of tort law. sarjain@stanford.edu

Susan Kenyon, Research Fellow in the Centre for Transport and Society, University of the West of England. Her research interests lie in the relationship between transport and social policy; the contribution of a lack of transport to the experience of social exclusion; in the impacts of new technologies upon transport and society; and in the efficacy of multidisciplinary research methodologies. Susan.Kenyon@ uwe.ac.uk

Christian Licoppe is Professor, Sociology of Information and Communication Technologies at the Department of Social Science, ENST, in Paris. Trained in History of Science and Technology, his current research focuses on understanding the regulation of mediated interactions and the accomplishment of IT-equipped activities in context, especially mobility and mobile phones, mediated consumption and call centres, the uses of the internet, interactions in professional settings. christian.licoppe@enst.fr

Stephen Little is Head of the Centre for Innovation, Knowledge and Enterprise, Open University Business School. He is currently working on the development of a Life Sciences version of the Open University MBA. He is researching new forms of mobility and regional development in the knowledge economy with institutions in the UK, Europe and Asia, and electronic governance in the UK and Africa. S.E.Little@open.ac.uk

Adrian Mackenzie, Institute for Cultural Research, Lancaster University, researches in the areas of technology, culture and social theory. He is currently preoccupied with the formations of agency, materiality and sociality associated with software (*Cutting Code: Software and Sociality*, 2005), and with understanding how infrastructures become more or less visible and contestable. a.mackenzie@lancaster.ac.uk

Ulrike Spring, is a Cultural Historian and Curator at Wien Museum, Vienna. She completed her D.Phil. on national language movements in Ireland and Norway at Vienna University in 2000. Her main research interests and publications concern national and hybrid identities, city and travelling in Europe from the eighteenth century onwards, focusing on medical reception and spatial perception. She has co-edited *Representing Gender, Nation and Ethnicity in Word and Image* (Kvinnforsk 2001) and has recently co-curated the exhibitions H.C. Anderson in Vienna (2005) and Mozart's Apartment (opening as part of the Mozarthaus Vienna in 2006).

David Murakami Wood is a Lecturer, Global Urban Research Unit (GURU) in the School of Architecture, Planning and Landscape at the University of Newcastle upon Tyne. He is the Managing Editor of *Surveillance & Society*. His research interests include: surveillance and social control, urban restructuring processes; Far Eastern cities; resilience of cities to contemporary threats; orbital space and vertical territoriality; actor-networks; and science fiction literature and film. d.f.j.wood@ ncl.ac.uk

CHAPTER ONE

Introduction: Mobile Cities, Urban Mobilities

Mimi Sheller and John Urry

Introduction

Cities are mobile places and places of mobility. Many cities were built at the convergence of major waterways and overland routes, and later became hubs for railways, highway systems, air transport and multi-modal metropolitan transportation systems. Mobility has been built into the infrastructure of cities, including the momentary immobilities of ports and freight depots, parking spaces and garages, airports and subway stops. 'Enlightened planners wanted the city in its very design to function like a healthy body, freely flowing . . . the Enlightenment planner made motion an end in itself' (Sennett 1994: 263–4). By the early twentieth-century the Chicago School urban theorists described mobility as 'perhaps the best index of the metabolism of a city' (Burgess 1925: 59). And today a 'new urbanism' emphasizes 'understanding cities as spatially open and cross-cut by many different kinds of mobilities, from flows of people to commodities to information' (Amin and Thrift 2002: 3); 'cities and urban regions become, in a sense, staging posts in the perpetual flux of infrastructurally mediated flow, movement and exchange' (Graham 2004: 154).

The constant flow of ships, ferries, horses, wagons, trains, trams, omnibuses, bikes, cars, planes, helicopters, underground metros and pedestrians has given the city its physical 'metabolism' (as interestingly shown in the case of nineteenth-century Singapore: Wong 2006). Cities also have an informational metabolism, having historically served as commercial hubs in which there was a density of informational flow concerning prices, arrivals and departures, availability of commodities and goods. Alongside this they functioned as centres of media and publicity with a density of meeting places, newspapers, leaflets, libraries, advertising posters, billboards and loudspeakers. Today such information flows, in part, take digital form, flowing through the dense urban infrastructure of coaxial and fibre optic cable, radio towers and satellite dishes, visual display units and, increasingly, WiFi protocols. Such flows – and the systems to control, channel and direct them – constitute cities as transnational entities made up of complex encounters, connections and mixtures of diverse hybrid networks of humans and animals, objects and information, commodities and waste (see Amin and Thrift 2002; Castells 1989, 1996).

Cities are also, of course, crucial sites for the arrival and departure of migrants, whether rural to urban movement within a nation-state, linear migration from one state to another, or the circular relocations and movements of 'transmigrants' and 'tramps' (Cresswell 2001). Diverse populations have long flowed into and out of cities, and continue to do so at a larger scale and with increasing rapidity. Tourism and business travel are also largely focused on cities, which serve as key arrival and departure points, meeting and conference centres, and tourist destinations in their own right (see many chapters in Sheller and Urry 2004). The attraction, management and surveillance of such flows and associated border controls has become one of the paramount concerns of urban governance and policing. And associated with the movements of populations come concerns about the movements of disease, from tuberculosis, HIV and SARS to foot-and-mouth, chicken flu and other disease vectors (see Little, this volume, and Law 2006). Fears generated by terrorism are also frequently associated with the mobility of people (and bombs) in and out of cities (as various papers in Sheller and Urry 2006a, show). Thus urban mobilities are not unlimited but are striated by 'a whole series of rules, conventions and institutions of regulation and control . . . a systematized network' (Amin and Thrift 2002: 26). Mobility is always 'differentiated' and ideologically enacted in processes of social, cultural and geographical power (Cresswell, forthcoming).

Urbanism, in sum, has always been associated with mobilities and their control, and continues to be so more than ever. The technologies, infrastructures, material fabric and representational machinery of cities support these mobilities, while also being shaped and re-shaped by them (see also Sheller and Urry 2000). As Stephen Graham has argued:

> the new hybrid interchanges of mobility and flow, as ICTs fuse with, and reconfigure, the other mobility spaces and systems of urban life, become critical and strategic sites at which the very political organization of space and society becomes continually remade.
>
> (2004: 155)

It is our contention in assembling this collection of essays on mobile technologies of the city that emergent practices of physical, informational and communicational mobility are reconfiguring patterns of movement, co-presence, social exclusion and security across many urban contexts. In particular, this book examines the interplay of physical mobility and mobile communications in the making of urban spaces, technoscapes and encounters.[1]

The aim of this volume is to bridge the gap between studies of physical mobility (e.g. transportation, migration, tourism studies) and informational mobility (e.g. the internet, media and mobile telephony), and in so doing to trace the emergence both of new socio-technical practices of the city and of a new theoretical paradigm for 'mobilities research'.[2] Many aspects of economic and social life are increasingly conducted 'on the move' or away from 'home'. In the convergences, conflicts and negotiations over physical mobilities and informational technologies urban futures will be determined. In a mobile world there are extensive and intricate connections between physical travel and modes of communication that seem to be forming new fluidities that are difficult to stabilize. Stabilizations of urban form are precarious, with new performances and representations of urban space often emergent, and with mobile communications technologies now assisting in the transformation of urban architectures, infrastructures, images and commodified value. Material changes in the technologies and infrastructures of urban life appear to be 'de-materializing', as connections, encounters and forms of co-presence are made and re-made in cities 'on the move'.

There are many studies of mobile communications and of urban transport systems, but very few that address the interplay between mobile communications, physical mobility and the city. A contemporary analysis of the relation between these systems, infrastructures and practices must be informed by recent theoretical developments in mobility studies, complexity studies, cultural geography, urban studies, cultural studies of technology, science studies and feminist studies of technoscience. These new theoretical approaches can contribute to re-thinking how emergent socio-technologies are re-shaping cities, urban form and space, while also enabling new forms of (im)mobility, subjectivity and encounter.

Much recent work on 'mobilities' tends to address a general condition of 'liquid modernity' (Bauman 2000) at an abstract theoretical level with little sense of the particularity of urban practices and the locatedness of mobilities at different scales (see Cresswell, forthcoming). In contrast, we address specifically located material practices as sites in which particular kinds of mobility and mobile communication have shaped or are re-shaping space, place and presence. Influenced by the material turn in European cultural geography and cultural sociology – which has made objects, infrastructures and physical 'stuff' an equal partner in the fabrication of socialities and technologies of co-presence – *Mobile Technologies of the City* aims to make mobilities, immobilities and their representations (in discursive, informational, visual and virtual form) central to contemporary urban studies. The specific case studies here involve Vienna, Liverpool, Bristol, London, Tokyo, Paris, Los Angeles and Hong Kong. They range across media including guidebooks, web sites, train schedules, WiFi, cinema, mobile phones and mobile gaming devices; and urban sites including streets, train platforms, bus stops, airports, internet cafés, park benches and cars. And they encompass infrastructures such as road and rail systems, airline networks and hubs, internet routers and wireless 'hot spots'.

3

The field of urban studies has mostly not addressed the impact of new technologies of mobile communication on reconstituting urban space (although see Amin and Thrift 2002; Crang 2000; Graham 1997, 2004; Moss and Townsend 1999; Townsend 2000). The strong and rich tradition of neighbourhood studies and place-based ethnography has in some ways limited the horizons of urban studies to what can be seen 'on the streets'. Deep immersion in a particular urban setting offers one perspective on urban life, but it also remains 'sedentarist' in its ontology (see Sheller and Urry 2006b), fixing urban space as a kind of bounded geographical location into and out of which people and objects move, but which has a kind of givenness. Instead, we need to ask: How are cities being de-materialized and re-materialized through new kinds of urban logics, technical systems and discursive orderings? How are new infrastructures transforming urban practices and the cultures of city life? How are new forms of virtual connectivity, including not only 'cyberpublics' but also innovations in mobile telephony and geo-located mobile gaming, reconfiguring urban encounters, and thus transforming the very 'stuff' of cities and citizenship? How are the pressing questions of security, violence, fear and terror that many urban polities face today being translated into new infrastructures of mobility, surveillance and selective immobility, or more precisely 'demobilization' (see Sheller 2004a; Verstraete 2003; Kaplan 2003; Graham and Marvin 2001; Graham 2004; Cresswell, forthcoming)?

Studies of new communication technologies in what is described as a 'wired world' offer one entrée into this subject, yet tend to ignore the physical mobilities of people and things, which intersect with the 'virtual' mobilities of communication. Dearnley and Feather's *The Wired World* (2001), for example, links the concept of the wired world to ideas of 'the information society' and the 'knowledge-based economy', and in doing so ignores questions of place, materiality and physical mobilities. Other edited collections have explored the emergence of 'networked societies' by focusing on 'cyberspace' and the internet. Armitage and Roberts' *Living with Cyberspace* (2002), for example, emphasizes new ways of understanding speed and space, but does not link this specifically to urbanism and questions of physical mobility. Many of the contributors to this volume, in contrast, pay close attention to the interplay of physical and informational mobilities and connectivity through studies of specific sites such as airports, mobile telephony devices, or wireless infrastructures. In doing so they also emphasize the importance of temporalities, and offer new perspectives on the making of place through the temporal coordination of mobilities.

Other strands within urban studies attend to specific developments in technologies of transportation, communication and surveillance within emergent 'digital cities' (e.g. Gandy 2002; Ishida and Isbister 2000; Tanabe *et al.* 2002; Bishop *et al.* 2003), but often fail to integrate these into a wider theoretical perspective on mobility and urbanism. Some work on 'CyberSociety' (Jones 1995) or 'Cyberdemocracy' (Saco 2002), includes an analysis of spatiality and community, but again without highlighting questions of complex mobilities/ immobilities. The work of Barry Wellman and his co-researchers has most consistently focused on the interplay between physical and virtual mobilities, focusing on the internet 'not as a special system but as routinely incorporated into everyday life' (Wellman and Haythornthwaite 2002: 6; see also Light 1999). The internet users most often addressed in these collections, however, are sedentary (either home-based or work-based) rather than the mobile users of various technologies (including both communicational and transportational) that is our focus in *Mobile Technologies of the City*. This volume thus offers one of the first empirically rich and theoretically informed explorations of mobile technologies that are only just coming into widespread use (see Funk 2004 for a more technical appraisal of the mobile internet market) and are beginning to restructure urban space and practice.

The field of transportation studies, where one might expect greater attention to questions of mobility and urban form, is in general not very theoretically informed (assuming a kind of positivist empiricism as common sense) and has missed out on many of the most interesting questions that this volume will address (see Sheller and Urry 2000, for a critique). As we have argued elsewhere, sociology's view of urban life failed to consider the overwhelming impact of the automobile in transforming the time-space 'scapes' of the modern urban/ suburban dweller. Industrial sociology, consumption studies, transportation studies and urban analyses have each been largely static, failing to consider how the car reconfigures urban life, with novel ways of dwelling, travelling and socializing in, and through, an automobilized time-space (see Sheller 2004b, and forthcoming). The socio-technical system of petroleum-based transportation infrastructure is not only a key form of contemporary mobility, but is, furthermore, interconnected with other mobile systems that organize flows of information, population, petroleum oil, risks and disasters, images and commodities (Sheller and Urry 2005, 2006).

Many mobility systems constitute and make possible cities. We turn, then, to a brief discussion of urban mobility as a system of intersecting systems, before giving an overview of the main lineaments of the book.

Urban Mobility Systems[3]

Through diverse studies of urban mobility systems we hope this volume will contribute to new theoretical perspectives on temporality, security and systematization. Each of the intersecting 'mobilities' that we have introduced above presupposes a 'system' (in fact, many such systems). These systems make possible movement and thus they make possible the city: they provide 'spaces of anticipation' that the journey can be made, that the message will get through, that the parcel will arrive. Centrally important here are systems that permit predictable and relatively risk-free repetition of the movement in question. In the contemporary city these systems include those of ticketing and licensing, oil and petroleum supply, electricity and water supply, addresses and postal systems, road safety and public safety, protocols, station interchanges, web sites, money transfer, luggage storage, air traffic control, barcodes, bridges, time-tables, CCTV surveillance and so on.

The history of these repetitive systems is, in effect, the history of those processes by which the natural world has been 'mastered' and made secure, regulated and relatively risk free. For people to be able to 'move', and for them, in turn, to move objects, texts, money, water, images, is to establish how it is that nature has been subdued (on nature, see Macnaghten and Urry 1998; Latour 2004). There is a metabolism that is effected by human societies over the physical world especially through developing and spreading diverse 'mobility-systems', and cities have been one of the central achievements of that mastery – a key symbol of both the pinnacles of civilization and the destruction of the natural world.

Historically many of these currently significant systems date from England and France in the 1840s and 1850s. Their interdependent development defines the contours of the modern mobilized world that brings about the 'mastery' of the physical world. In mid-nineteenth-century Europe nature gets dramatically and systematically 'mobilized'. Systems dating from that exceptional moment include a national post system (the Penny Post), the invention of photography and their use within guide books (Daguerre in France, Fox Talbot in England), the first railway age and the first ever national railway timetable (Bradshaws), the first city built for the tourist gaze (Paris), the first inclusive or 'package' tour (organized by Thomas Cook), the first scheduled ocean steamship service (Cunard), the first railway hotel (York), the early department stores (in Paris), the first system for the separate circulation of water and sewage (Chadwick in Britain) and so on. Across the colonial world in the mid- to late nineteenth century people also witnessed (and contributed labour to) the building of roads, canals, railways, ports and systems for the regional and world-wide shipment of goods and people, usually via burgeoning colonial port cities. Morse code, the telegraph and then the telephone were developed during the latter years of that century.

The twentieth century then of course saw a huge array of other 'mobility-systems' develop, including the car system, national telephone system, air power, high-speed trains, modern urban systems, cheap air travel, mobile phones, networked computers and so on. Virilio maintains that systems are increasingly developed in which there is an obligation to be circulating, and this is true of water, sewage, people, money, ideas (1986). Circulation is a powerful notion here that has many impacts upon the social world. There is in the modern world an accumulation of movement analogous to the accumulation of capital – accu-mulations of repetitive movement or circulation made possible by diverse, interdependent mobility systems (see Thrift on 'movement-space', 2004) And as we move into the twenty-first century these 'mobility systems' are developing some novel characteristics, traced out more fully in Urry (2006).

First, such mobility systems are more complicated, made up of many more elements and based upon an array of specialized and arcane forms of expertise. Mobilities have always involved expert systems but these are now highly specific, many being based upon entire university degree programmes and specialized companies. Second, such systems are much more interdependent with each other so that individual journeys or pieces of communication depend upon multiple systems, all needing to function and interface effectively with each other. Third, since the 1970s, systems are much more dependent upon computers and software. Software we might say, paraphrasing Thrift (2001), writes mobility. There has been a massive generation of specific software systems that need to speak to each other in order that particular mobilities and 'sortings' take place. Pervasive computing produces a switching and mobility between different self-reproducing systems, such as the internet with its massive search engines, databases of information storage and retrieval, world money flows especially through the ubiquitous 'spreadsheet culture', intelligent transport systems, surveillance systems and so on. Fourth, these systems have become especially vulnerable to what Perrow terms 'normal accidents', accidents that are almost certain to occur from time to time, given the tightly locked-in and mobile nature of many such interdependent systems (1999; see Law 2006).

Such increasingly complex, computerized and risky systems are central to new urban forms and practices, and to the lives of city dwellers. As daily and weekly time-space patterns in the richer parts of the world are desynchronized from historical communities and place, so systems provide the means by which work and social life can get scheduled and rescheduled. Organizing 'co-presence' with key others (workmates, family, significant others, friends) within each day, week, year and so on becomes more demanding with this loss of collective coordination. 'Clusters' dissolve into what Wellman terms 'personalized networking', a person-to-person connectivity most revealed now with those machines that enable immediate, mobile connectivity (2001). The greater the personalization of networks, the more important are systems to facilitate that personalization. There is a spiralling, adaptive relationship effected through 'scheduling systems', while of course much of the world's population are unable to participate in a life on the move and are thereby more socially excluded; or they are subjected to movements against their will, or else find their mobilities involuntarily delimited by passports and visas, border-guards and detention centres, and what some call the 'prison-industrial complex' (Sudbury 2004; and see Ahmed *et al.* 2003).

With de-synchronization the use of scheduling becomes more necessary. There is an increasingly 'do-it-yourself' scheduling society commonplace in at least large cities across the world. And the greater the personalization of networks, the more important are systems to facilitate that personalization. There are irreversible changes taking place that are moving social connections towards person-to-person networks requiring specific personalized scheduling systems in order for life on the intermittent move to take place. And those systems are especially necessary as system risks and failures abound and arrangements need to be endlessly renegotiated. Networks are on occasion tightly coupled with complex, enduring and predictable connections between peoples, objects and technologies across multiple and distant spaces and times (Murdoch 1995: 745;

Law 1994: 24). Things are made *close* through these networked relations, and both *closeness* and *distance* are made or unmade, rather than being simple objective measures of geographical placement.

What implications do these changes have for the wider fabric of urban infrastructure and urban life? Graham and Marvin's *Splintering Urbanism* (2001) especially pays attention to questions of exclusion, disconnection, bypassing and differentiation that are central to thinking about mobilities and their implications (and are further explored by Wood and Graham in this volume). Crucial here are questions of civic participation, social exclusion and the transformation of urban public spaces, which we explore below in terms of transformations in the systems supporting co-presence and mobilities. Personalization, ineluctably, comes with systems of sorting, tiering, channelling, and differentiating access for 'preferred customers'. Third-generation wireless communication systems in the US, for example, claim to offer 'ubiquitous' and 'instantaneous' communications for business customers moving across a world of seamless high-speed connections enabled by expensive devices (Schiesel 2005), but such premium services are predicated on the dismantling of the ideals and infrastructures of universal access that once underpinned the public utilities and the social inclusion of the entire 'public' in urban public space.

New urban transport infrastructures are also being built and planned mainly to better service the business customer or to stimulate corporate investment in specific urban areas while bypassing others. From the Docklands light railway and successful congestion charging scheme in London, to huge infrastructural developments in Malaysia, Singapore and China (Graham and Marvin 2001), large-scale, new urban transportation projects have been promoted as good for business, good for tourism, and good for increasing circulation through congested city centres. In New York various interest groups and speculators are jockeying for position as multiple plans are floated to redevelop the urban transport infrastructure.[4] All of these potential projects are linked to visions of the informational infrastructure of the future, to projections of the desirability of increased 'mobility' as a public good, and to media representations of ideals of urban mobility – even when their realization will inevitably exclude, bypass, and lead to disinvestment in other geographical areas.

In many ways, then, the reconfiguration of complex mobility and communication systems is not simply about infrastructures but the refiguring of the public itself – its meanings, its spaces, its capacities for self-organization and political mobilization, and its multiple and fluid forms (see Sheller and Urry 2003; Sheller 2004c). In the following section we introduce some further key concepts and provide an overview of the sections and chapters of the book.

Overview of the Book

We employ the notion of 'technoscapes' to explore emergent socio-technologies and their spatial forms. Technoscapes is a term drawn from the work of Arjun

Appadurai and also John Urry's more general concept of 'scapes'. As it is used here it builds on the notion of 'landscape' as a simultaneous representation *and* shaping of land and space. We can think of landscape as a combination of social scripts, cultural values, and material rearrangements of physical elements (trees, lakes, buildings, walkways, canals, etc.) that organizes views, viewers, space and the elements within it (vistas, viewpoints, landmarks, etc.). Subjects move through a socially structured landscape and thus help to perform that landscape, but landscapes also afford or perform certain kinds of subjectivities. Landscapes are imbued with power, enabling particular forms of domination and subordination marked out in sites such as central squares, military fortifications, imposing façades and gates, urban skyscrapers or rural landholdings marked out with parkland, tree-lined drives or architectural landmarks.

Technologies also work in this way, and we describe this work as a 'technoscape'. The concept of technoscape serves to emphasize that contemporary landscapes are shot through with technological elements which enrol people, space, and the elements connecting people and spaces, into sociotechnical assemblages – especially the transportational technologies, such as roads, rail, subways and airports, but also the informational technologies such as signs, schedules, surveillance systems, radio signals and mobile telephony cells. People move through these technoscapes whenever they ride in a train, make a phone call, read a computer screen, simply step off a pavement to cross a road, or hike on a marked trail. Data, pictures and sounds also flow through technoscapes (implying both an infrastructure and a set of practices through which people access such flows).

A technoscape is also associated with a range of equipment that enables the production and consumption of space: a car, a mobile phone, a camera, a screen, a hiking boot and so on. Law argues that 'what we call the social is materially heterogeneous: talk, bodies, texts, machines, architectures, all of these and many more are implicated in and perform the social' (Law 1994: 2; and see Law 2006). From science and technology studies we take the idea that architectures, machines and texts enable or 'afford' the possibility for certain kinds of mobilities and immobilities. Human, non-human and inhuman agents interact via the affordances of the spaces, infrastructures and technologies in and through which they move, pause, dwell and encounter one another. 'The rhythms and motions of these inter-corporeal practices', suggests Whatmore, 'configure spaces of connectivity between more-than-human life worlds; topologies of intimacy and affectivity that confound conventional cartographies of distance and proximity, and local and global scales' (Whatmore 2002: 162; and see Clark 2001 on 'trans-human' interchanges in the metropolis).

Like a landscape, a technoscape also performatively produces particular subjectivities (such as 'the driver', 'the pedestrian', 'the hiker', etc.). Several of the contributors here (Spring, S. Jain, Mackenzie, Licoppe and Guillot) carefully investigate not only the technologies of mobility, but also the technologies of imagining, fantasizing and representing mobility as a desired good, and in particular how these imaginaries are linked to urbanism. 'People and places script each other', argue Amin and Thrift (2002: 23), and 'cities are intensely visualized through images of one sort or another', including tourist maps, city guides, aestheticized districts, architectural signatures, media and film. Through these conglomerations of communication, representation and naming, places enrol, arrange and include or exclude people, while people perform places through imaginaries.

The term 'imaginaries' refers to 'the French idea of the imaginary (*imaginaire*), as a constructed landscape of collective representations, which is no more and no less real than the collective representations of Emile Durkheim, now mediated through the complex prism of modern media' (Appadurai 1996: 73). Accordingly, Ulrike Spring (this volume) shows how the city guidebook shaped the urban form of nineteenth-century Vienna, acting not only as a representation of the city and a script for moving through the city, but also as a kind of blueprint for urban design. And Adrian Mackenzie (this volume) calls for an 'urban new media studies' that would 'analyse the mutual contextualization of images of movement and movement itself, particularly when movement itself becomes an image'. The technoscape and the mediascape therefore work together to produce urban forms, urban imaginaries and urban subjects of particular kinds.

The contemporary technoscape has its own genres of visual representation, especially the cinematic narrative and the commercial advertisement, which are deeply co-constituted with and by the technologies of mobility and communication. Sarah Jain (this volume) argues in an analysis of BMW advertising, that 'how the car works as a commodity, and particularly through its materiality as a *mobile* commodity, is necessary to understanding the urban form'. Urban form (especially in Los Angeles) is shaped by the car's violent 'taking of space', she argues, and this is celebrated and perpetuated by the celebrity culture attached to luxury commodities such as the BMW. Christian Licoppe and Romain Guillot (this volume) further observe how the representation of futuristic urbanism in films such as *Blade Runner* and *The Matrix* (whose urban aesthetic is based in part on districts of modern Tokyo) influenced the design of mobile games, and the playing of these games in turn has the capacity to transform urban space and forms of encounter in cities such as Tokyo.

Thus, the interplay between 'the real' and 'the virtual' involves complex interactions and feedback effects. First, representations of mobility and iconic mobile commodities influence the actual urban form, aesthetically and kinaesthetically, in terms of their design, form and capabilities; second, movements through space are scripted to perform urban space according to the dominant genres, rules, architectures and infrastructures; but, third, on-the-ground implementations of the new technologies of mobility and communication nevertheless have the potential to transform cities both through their power to re-make or re-deploy visual representations of urban form and through their inadvertent openness to the improvisations by which people enact the city. Designers, likewise, must create mobile artefacts that generate profit for business through their desirability and use, but they must also recursively respond to the actual ways in which people use the objects and software that they design (Licoppe and Guillot, this volume). Mobility systems (and hence cities) are therefore always dynamic, emergent and open to invention despite being rooted in the cultures of capital, commodification, surveillance and control.

Technologies do not produce necessary effects (i.e. technological determinism), but are part of the bundle of actants through which agency is scripted, produced, enacted, contested and repeated. Indeed, new transport technologies are often very slow in their uptake (see Pooley *et al.* 2006), while in other cases the accumulation of small repetitions – as with the growth of mobile phone use or communications between offices using faxes – can lead to a 'tipping point' in which an entire system is transformed, it seems, overnight (Gladwell 2000). Thus, the contributors to this volume pay close attention to the mundane ways people actually use technologies, the manners in which designers implement and adapt devices and systems, and how infrastructures are culturally as well as technically embedded. As Amin and Thrift put it: 'Each urban moment can spark performative improvisations which are unforeseen and unforeseeable' (2002: 4).

The chapters in Part I, Mobilities and the Creation of Urban Spatial Form, offer sophisticated cultural analyses of the creation of urban form in relation to fantasy, representation and commodification. This section underlines the crucial role of mobilities in the framing of urban space, and the ways in which space and mobility intersect with narrative and imaginary representations of the city. Concerns with the interrelation of representation and spatiality are not new to urban studies, dating at least to Walter Benjamin, but the contributors here pay special attention to the generative effects of technologies of mobility (guide books, airports, cars) as they collude in representational practices (city images, web sites, film) and in the service of cultures of enterprise (tourism, internet start-up businesses, advertising).

In Chapter 2 Ulrike Spring considers how guidebooks and sightseeing tours introduced new ways of experiencing the city of Vienna in the nineteenth century, and shows how new practices of walking through the city were interrelated with material changes in the urban form. Infrastructural improvements, new kinds of public spaces and new urban spatial narratives paved the way for what is now contemporary mass tourism. This chapter illustrates how novel approaches to mobility study can be applied to historical cultural studies as well as to contemporary analyses.

In Chapter 3 Peter Adey and Paul Bevan examine how space has become reorganized, recombined and permeated by technologies of extended virtual connectivity (producing what they call 'technospaces'). They use empirical case studies of an internet web site and a regional airport, both based in the Liverpool (UK) area, to explore the connectivity of physical and virtual 'cyber-mobilities'. This chapter offers a well-grounded empirical analysis of two key sites (airports and the internet) that are central to contemporary claims about changing mobilities.

And in Chapter 4 Sarah Jain analyses the co-constitutive relation between violence, automobiles and the US American urban form, including the place of fantasy (in the form of a short film directed by Guy Ritchie, starring Madonna, and appearing on the BMW web site) in promoting the car as a vehicle for making and taking space. Jain argues that an understanding of the materiality of the car as a mobile commodity is necessary to understanding contemporary urban form, and the violence that underlies it. With a US-focus and a high-profile star as its subject, this chapter brings mobility studies into conversation with media studies and a sophisticated cultural political economy.

The chapters in Part II, Re-configuring Co-presence, examine how new configurations and coordinations of physical mobility with communications technologies are producing new social practices of travel and new system-atizations, along with changing exclusions and risks. These case studies suggest some of the ways in which the field of transport studies is being challenged by new informational mobilities and their effects on time-space planning, predic-tion and risk. Co-presence refers to face-to-face or in-the-same-place encounters of a person with another person, a place, an object or an event. The terms 'virtual reality' or 'cyberspace' are often used to refer to mediated non-physical presences, but they carry many implications and valuations about the status of 'reality' and 'real space' versus 'simulacra' and 'simulations'. 'Co-presence' is meant as a more neutral description of encounters in the physical world, without any implicit assumptions about the ontological difference between the real, the imaginary, the virtual, the simulated, etc., since it is precisely their relation which some chapters in this book will explore. Insofar as 'the actual and the virtual constantly inscribe each other' (Amin and Thrift 2002: 119) how do these inscriptions channel mobilities and immobilities?

In Chapter 5 Juliet Jain draws on research undertaken with a variety of transport organizations to examine how the ability to deliver personalized travel information to people on the move is reconfiguring mobility practices involving the combined use of mobile technologies and existing infrastructures of the British national rail network. She brings to the fore issues of timing and sched-uling, which are central to theorizing urban mobilities. Attention to a public transport system helps to counter the overwhelming tendency to focus on indi-vidualized representations and accounts of 'business class' mobility in the commercial sector and mass media.

Chapter 6 is concerned with methodologies for assessing the impact of virtual mobility, via the internet, upon personal travel and social participation. Susan Kenyon shows how 'accessibility diaries' (used in a study in the south-west of England) might be used to explore issues of mobility-related social exclusion. This chapter contributes to debates about the interaction between virtual mobility and physical mobility not only by devising a robust empirical instrument for exploring the 'substitution' hypothesis, but also by connecting 'accessibility' to issues of social exclusion. It is also relevant to the develop-ment of more 'mobile methods' to investigate the new questions being raised by mobilities research (see Sheller and Urry 2006b).

Chapter 7 focuses on two events that replaced predictability with uncertainty and increased the perceived risk of international air travel: the 11 September attacks on the US in 2001 and the 2003 SARS outbreak in East Asia. These two events, Stephen Little argues, changed collective understandings of the relation between place, transport and accessibility, and underpinned new forms of surveillance. This chapter highlights how these two crucial global events affected transport policy both internationally and locally, and repositioned issues of mobility, risk and complexity.

The chapters in Part III, Cultures of Infrastructure and Public Space, draw on theoretical tools inspired by actor-network theory and social studies of technology, but are also closely grounded in empirical research on the implementation and development of specific new technologies. Each chapter breaks new ground in offering original empirical research that challenges theoretical presuppositions and advances new approaches to urban mobility studies. By problematizing that which we take for granted or ignore as background (everyday infrastructures, encounters in public spaces, and software systems), the chapters in this section draw attention to the mundane yet powerful forces shaping contemporary urbanism. Like some of the earlier chapters, they also address ethical questions of the uses to which new socio-technical systems are put and, in particular, the role of designers in generating unintended consequences or unforeseen uses of devices and software.

In Chapter 8 Adrian Mackenzie complicates the idea of infrastructure by suggesting the ways in which new technologies are produced as much by images and cultural practices as they are by designers, engineers, builders and planners. He shows how the emerging figurations, practices, commodifications and modifications associated with WiFi involve a repositioning of people in relation to the movement of data, as well as fomenting a powerful image of movement. Mobile technologies are 'heavily imagined' such that any analysis of mobility must also be an analysis of 'images of movement'. Thus he calls for an 'urban new media studies' that would work at the interface of urban infrastructures, new media technologies, processes of design and commodification, and the imagination of mobility.

In Chapter 9 Christian Licoppe and Romain Guillot explore how the interpretation and use of a new mobile technology is emergent and depends on the particular organization of the situations in which it is used. Using participant observation and interviews with a Parisian startup mobile game-designing firm, as well as game-users in the Tokyo-Yokohama conurbation, this chapter shows how a design team's plans for a multi-player role playing game shifted from an initial emphasis on coordinated team movements through the city and on screen, to become instead an innovative but individualized geo-located information and communication technology that was used in unforeseen ways. They show how both designers and game-users interact with urban spaces and demonstrate that through the development of location-based services the very notion of encounters and social order in public spaces is being transformed.

Chapter 10, finally, also draws on actor-network theory and the sociology of technology to examine automated systems of surveillance, which are part of the 'new social control' of a society in which software is used to sort categories of highly differentiated mobility. David Marakami Wood and Stephen Graham argue that software applications (such as iris-scanning technology in airports or internet routers on the World Wide Web) are being used as powerful 'sorting' devices to generate high- and low-mobility corridors and to 'unbundle' public spaces and services, leading to the progressive splintering of previously largely integrated public domains, infrastructures and spaces of cities.

Further Research

Having drawn the contours of this book, it remains to point to some of its limitations. Above all, further research is needed into the uneven distribution of mobilities across stratified formations of gender, race, ethnicity, class and sexuality. Mobilities are arguably crucial to the marking of racial and ethnic boundaries and to the making and unmaking of colonial and postcolonial gender and racial orders (Sheller 2003). The operation of national boundaries, borders and identities works to signify where one population group ends and another begins, and recent attention to 'borderlands', 'border crossing' and 'border identities' demarcates a crucial terrain within mobilities studies (Ahmed *et al.* 2003). While much work on borders focuses on the national border, imagined as an edge on a map, the study of urban mobility systems requires greater attention to the making and crossing of borders in the micro-processes of the urban technoscape – e.g. the uses of the identity card and biometric recognition, the policing of illegal immigrants and terrorist suspects, the rounding up of 'trafficked' women and children – and the ways in which racial, ethnic and gender identities get made and re-made at such borders.

In relation to gender we know that men and women have had very different access to the means of mobility and communication in many parts of the world and throughout history (see Cresswell forthcoming). We also know that the discourses of mobility and travel are deeply informed by the projects of Western masculinity, in particular a kind of masculine bourgeois individualism that promotes autonomy, freedom and cosmopolitanism (see Sheller and Urry 2006b). This context needs to be brought into analyses of the forms of women's urban mobility, especially the large body of research on sex work, trafficking, mail-order brides, sex tourism and urban prostitution. In so far as urban spaces are both gendered and sexualized, how do both histories of mobility and changing forms of physical and informational mobility contribute to the production, reproduction, or transformation of such patterns?

Finally, we could suggest that 'urban culture' as a whole is a product of mobilities. Cities are very much the crucibles of cultural juxtaposition, fusion, hybridization and syncretism. The ethic of urbanism is to foster circulations of cultural creativity, invention and innovation, which depend on the many systems of urban mobility that we have outlined above. A kind of dynamic of cultural movement and invention – sometimes called creolization – is at the heart of mobile cities and urban mobilities. We would therefore call for a more mobile urban studies which also attends to cultural difference, to global flows, and to the dynamic relations between historical metropoles and peripheries as well as translateral linkages across less central sites; one that acknowledges not only the convergences and interferences of new transport infrastructures and new mobile communication systems in the major 'global cities', but also explores the effects of these interacting mobilities on constituting diverse subjects and cultures of urbanism around the world.

Notes

1   This volume arose out of the Alternative Mobility Futures conference organized by the Centre for Mobilities Research at Lancaster University (January, 2004). We wish to thank the Institute for Advanced Studies in Management and Social Science for supporting this conference, and participants in the conference for their excellent contributions and discussions. Also see Sheller and Urry 2006a, for some further papers given at this conference. Mimi Sheller and John Urry are immensely grateful to Pennie Drinkall for her assistance in compiling this complex book for publication and also for organizing the conference.
2   There have been numerous studies of the macro-economic and macro-spatial relation between urban mobilities and information and communication entrepreneurship in terms of the new information economy or knowledge-based economies of 'global cities' (Sassen 2001) and the emergence of 'network society' (Castells 1996). However, our aim here is to focus first on the 'lived' spaces of everyday urbanism and the actual ways in which people navigate physical, digital and virtual terrains in 'real-time', and second, on the representational realms in and by which urban mobilities are imagined, scripted and enacted.
3   This section is developed in more detail in the forthcoming Urry 2006.
4   Options on the table at the moment include rebuilding a Lower Manhattan transport hub in the wake of the World Trade Center attack; tunnelling under the East River to create a new connection from Wall Street to the Kennedy International Airport; expanding Penn Station; extending the No. 7 subway line to the Far West Side; building a freight-train tunnel under New York Harbour; and building a $5 billion rail tunnel under the Hudson River to help New Jersey employees commute to mid-town Manhattan companies (McGeehan 2005).

Works Cited

Ahmed, S, Castañeda, C, Fortier, A-M and Sheller, M (eds) (2003) *Uprootings/ Regroundings: questions of home and migration*, Oxford and New York: Berg.
Amin, A and Thrift, N (2002) *Cities. Reimagining the Urban*, Cambridge: Polity.
Appadurai, A (1996) *Modernity at Large: cultural dimensions of globalization*, Minneapolis: University of Minnesota Press.
Armitage, J and Roberts, J (eds) (2002) *Living with Cyberspace: technology and society in the twenty-first century*, New York and London: Continuum.
Bauman, Z (2000) *Liquid Modernity*, Cambridge: Polity.
Bishop, R, Phillips, J and Yo, W (eds) (2003) *Perpetuating Cities: postcolonial urbanism in South East Asia*, London and New York: Routledge.

Brendon, P (1991) *Thomas Cook: 150 Years of popular tourism*, London: Secker and Warburg.

Burgess, E (1925) [1970 edn] 'The growth of the city: an introduction to a research project', in R Park, E Burgess and R McKenzie (eds) *The City*, Chicago, IL and London: University of Chicago Press.

Castells, M (1989) *The Informational City*, Oxford: Blackwell.

—— (1996) *The Rise of the Network Society*, Oxford: Blackwell.

Clark, N (2001) '"Botanizing on the asphalt"? The complex life of cosmopolitan bodies', in P Macnaghten and J Urry (eds) *Bodies of Nature*, London: Sage.

Crang, M (2000) 'Public space, urban space and electronic space', *Urban Studies*, 37, 2: 301–17.

Cresswell, T (2001) *The Tramp in America*, New York: Reaktion Books.

—— (forthcoming) *On the Move: the politics of mobility in the modern West*, New York: Routledge.

Dearnley, J and Feather, J (2001) *The Wired World: an introduction to the theory and practice of the information society*, London: Library Association.

Funk, J (2004) *Mobile Disruption: the technologies and applications driving the mobile internet*, Hoboken, NJ: Wiley Interscience.

Gandy (2002) *Concrete and Clay: reworking nature in New York City*, Cambridge, MA: MIT Press.

Gladwell, M (2000) *Tipping Points. How little things can make a big difference*, Boston: Little, Brown and Company.

Graham, S (1997) 'Imagining the real-time city: telecommunications, urban paradigms and the future of cities', in S Westwood and J Williams (eds) *Imagining Cities*, London: Routledge.

—— (ed.) (2004) *The Cybercities Reader*, London: Routledge.

—— and Marvin, S (2001) *Splintering Urbanism*, London and New York: Routledge.

Ishida, T and Isbister, K (eds) (2000) *Digital Cities: technologies, experiences and future perspectives*, Berlin: Springer-Verlag.

Jones, S (ed.) (1995) *CyberSociety: computer-mediated communication and community*, Thousand Oaks, CA: Sage.

Kaplan, C (2003) 'Transporting the subject: technologies of mobility and location in an era of globalization', in S Ahmed, C Castañeda, A-M Fortier and M Sheller (eds) *Uprootings/Regroundings: questions of home and migration*, Oxford and New York: Berg.

Latour, B (2004) *Politics of Nature*, Cambridge, MA: Harvard University Press.

Law, J (1994) *Organizing Modernity*, Oxford: Basil Blackwell.

—— (2006) 'Disaster in agriculture: or foot and mouth mobilities', *Environment and Planning A*, (in press).

Light, J (1999) 'From city space to cyberspace', in M Crang, P Crang and J May (eds) *Virtual Geographies: bodies, spaces, relations*, New York: Routledge, pp. 109–30.

McGeehan, P (2005) 'New in the cellar at Macy's: a tunnel to New Jersey?', in *The New York Times*, Metro Section, 13 January 2005, p. B1/B6.

Macnaghten, P and Urry, J (1998) *Contested Natures*, London: Sage.

Moss, M and Townsend, A (1999) 'How telecommunications systems are transforming urban spaces', in J Wheeler and Y Aoyama (eds) *Fractured Geographies: cities in the telecommunications age*, New York: Routledge.

Murdoch, J (1995) 'Actor-networks and the evolution of economic forms: combining description and explanation in theories of regulation, flexible specialisation, and networks', *Environment and Planning A*, 27: 731–57.

Perrow, C (1999) *Normal Accidents*, Princeton, NJ: Princeton University Press.

Pooley, C, Turnbull, J and Adams, M (2006) 'The impact of new transport technologies on intra-urban mobility: a view from the past', *Environment and Planning A* (in press).

Saco, D (2002) *Cybering Democracy: public space and the internet*, Minneapolis, MN and London: University of Minnesota Press.

Sassen, S (2001) *The Global City*, 2nd edn, Princeton, NJ: Princeton University Press.

Schiesel, S (2005) 'For wireless, the beginnings of a breakout', *The New York Times*, Circuits Section, Thursday, 13 January 2005 p. G1/G5.

Sennett, R (1994) *Flesh and Stone*, New York: Norton.

Sheller, M (2003) *Consuming the Caribbean*, New York and London: Routledge.

—— (2004a) 'Demobilizing and remobilizing Caribbean paradise', in M Sheller and J Urry (eds) *Tourism Mobilities: places to play, places in play*, London and New York: Routledge, pp. 13–21.

—— (2004b) 'Automotive emotions: feeling the car', *Theory, Culture and Society*, 21, 4/5: 221–42.

—— (2004c) 'Mobile publics: beyond the network perspective', *Environment and Planning D: society and space*, 22: 39–52.

—— (forthcoming) 'Bodies, cybercars and automated mobilities', *Social and Cultural Geography*, Special Issue on Banal Mobilities: Performances, Practices, Materialities.

Sheller, M and Urry, J (2000) 'The city and the car', *International Journal of Urban and Regional Research*, 24, 4: 737–57.

—— (2003) 'Mobile transformations of "public" and "private" life', *Theory, Culture and Society*, 20, 3: 107–25.

—— (eds) (2004) *Tourism Mobilities: places to play, places in play*, London: Routledge.

—— (eds) (2006a) *Mobilities and Materialities*. Special issue of *Environment and Planning A* (in press).

—— (2006b) 'The new mobilities paradigm', *Environment and Planning A* (in press).

Sudbury, J (ed.) (2004) *Global Lockdown: race, gender and the prison-industrial complex*, London: Routledge.

Tanabe, M, ven den Besselaar, P and Ishida, T (eds) (2002) *Digital Cities II: computational and sociological approaches*, Berlin: Springer-Verlag.

Thrift, N (2001) 'The machine in the ghost: software writing cities', paper presented to Hegemonies Conference, Lancaster: Centre for Science Studies, Lancaster University, September.

—— (2004) 'Movement-Space: the development of new kinds of spatial awareness', paper presented to Alternative Mobility Futures Conference, Lancaster: Centre for Mobilities Research, January.

Townsend, A (2000) 'Life in the real-time city: mobile telephones and urban metabolism', *Journal of Urban Technology*, 7: 85–109.

Urry, J (2006) *Mobilities*, Cambridge: Polity.

Verstraete, G (2003) 'Technological frontiers and the politics of mobility in the European Union', in S Ahmed, C Castañeda, A-M Fortier and M Sheller (eds) *Uprootings/ Regroundings: questions of home and migration*, Oxford and New York: Berg.

Virilio, P (1986) *Speed and Politics*, New York: Semiotext(e).

Wellman, B (2001) 'Physical place and cyber place: the rise of networked individualism', *International Journal of Urban and Regional Research*, 25: 227–52.

—— and Haythornthwaite, C (eds) (2002) *The Internet in Everyday Life*, Malden, MA: Blackwell.

Whatmore, S (2002) *Hybrid Geographies: natures, cultures, spaces*, London: Sage.

Wong, Y-S (2006) 'When there are no pagodas on Pagoda Street: language, mapping and navigating ambiguities in colonial Singapore', *Environment and Planning A* (in press).

PART I

Mobilities and the
Creation of Urban
Spatial Form

CHAPTER TWO

The Linear City: Touring Vienna in the Nineteenth Century

Ulrike Spring

Linearity exposed

A traveller visiting Vienna in the early and the late parts of the nineteenth century respectively, encountered a very different city indeed. Having in many respects turned into a modern city, Vienna now possessed an expanded area, widened streets and numerous splendid buildings. Moving through the city had become easier and the transitory and fragmentary character of urban existence would in many cases have been overlaid by lines of traffic or paths of tourism. This change in the material infrastructure and forms of mobility corresponded to new urban imaginaries. Indeed, as I would like to argue, a linear perception slowly asserted itself in the course of the nineteenth century not just in Vienna, but in other European cities as well, and moreover this has had a significant impact on tourists' perception of our own time.

With this, I do not intend to further the impression that the linear city might comprehend every aspect of urban life and perception. Rather, one could consider the linear city, applying Donald's (1999: 8) terminology, as a variant of an imagined environment:

> This environment embraces not just the cities created by the 'wagging tongues' of architects, planners and builders, sociologists and novelists, poets and politicians, but also the translation of the places they have made into the imaginary reality of our mental life.

Not only is the linear city an integral part of urban life, it is also what, together with other city images, shapes and produces urban life. It constitutes part of this imagined city, while at the same time merging with other forms of urban imaginary. Moreover, the linear city is (re)produced at the intersection of local and global perceptions, knowledge and representations. It emerges in the act of the traveller reading a guidebook, of him or her listening to urban narratives or memories; but it also occurs as a result of the traveller meeting with the materiality of the city, with its streets, transportation and sights. Hence, the urban experiences of the traveller and of the local inhabitant meet at certain points, and diverge at others. To a certain extent, the city absorbs the traveller into its social space or mental topography.

Certainly, the linear city cannot be called an as yet 'undiscovered city' (Thrift 2000). In architecture, the term rose to prominence with Arturo Soria y Mata and his 'ciudad lineal'-movement, and above all with Le Corbusier's utopian (or dystopian, depending on how one sees it) visions of the regulated city, constituted by grids and linked by lines of traffic. While here the idea of linearity is concretized in the material body of the city, the development of a linear perception in the eighteenth and nineteenth centuries has also been noted in other contexts, such as in painting (Barrell 1972), theories of national narratives tracing lines from mythic origin to future glory (Smith 1991), the natural sciences with their concept of evolution and, more generally, that of velocity (Virilio 1984). However, my focus will be on a topic that has not been so widely explored: sightseeing tours and their interrelatedness with material changes in the city, which I argue reproduced and affirmed linear perception in mental life. This linear perception involves improvements of infrastructure, new practices of social control and agency, and last but not least the assertion of new forms of travel, laying the path for mass tourism.

More specifically, I want to suggest that the mapping of tours upon urban space indicated a new way in which the city was to be consumed by tourists. Indeed, the introduction of sightseeing tours as a common mode of experiencing the city represented, in part at least, a new way of approaching urban spatial narratives, characteristic of city travel up until the present day. Furthermore, I will argue that this has to do with a change in patterns of mobility that had already emerged in the eighteenth century, but only became widespread in the course of the nineteenth century. Although I will focus on the modes of perception guidebooks provided for visitors, my thesis is based on the assumption that urban spatial perception always emerges at the intersection of local and foreign experiences and cultural representations.

While this change in the perception of urban space may be observed in many European cities in the course of the nineteenth century, Vienna will serve as my case-study, first because of its geopolitical position as an imperial Central-European metropolis and hence tourist attraction in the nineteenth century, and second because of its status as a city caught in ambivalence between modernization and conservation in the late nineteenth century.

Tracing Lines: Sightseeing Tours in Vienna

Vienna underwent major changes in its material structure from the late 1850s onwards. In 1858 the extension of the city within the framework of urban planning was to begin, having as its explicit aim the building of a capital worthy of its role as imperial metropolis (Breitling 1980: 35). Although some contemporaries considered the great architectural changes in Vienna in the second half of the nineteenth century as equal to Haussman's projects in Paris, they never reached the same dimensions. Yet the implementation of the new urban policies in Paris in the second half of the nineteenth century led to 'a new way of imagining the city', one that became explicit in the turn-of-the-century politics of urban planning (Donald 1999: 27), and Vienna was drawn into this process as well. According to Breitling (1980: 41), the Vienna planning competition of 1858, preceding the extension of the city, 'marked the end of an era, and in a sense ushered in the new urban age of the second half of the nineteenth century'. However, the process of transforming Vienna into a metropolis on a similar footing to Berlin or Paris did not run smoothly and, in fact, did not seriously take root before the 1890s (see Meißl 2000: 286). At this time, a new public transport system was built (the city train or Stadtbahn), gas and electricity supplies were extended, and heated debates arose around the advantages of either changing Vienna into a modernized city or preserving its old style (Meißl 2000). The perspectives and expectations of travellers to Vienna were accorded a significant role in these discussions. As the *Verein zum Schutze und zur Erhaltung der Kunstdenkmäler* (Society for the Protection and Preservation of Artistic Monuments), highly critical of Vienna's modernization, wrote in 1909, the focus on the impressive Ringstraße often neglected 'what for foreigners, in particular Americans . . . possesses the greatest attraction, being the old quarters of Vienna with their narrow and winding streets'[1] (*Neues Wiener Tagblatt*, 13.4.1909, quoted in Meißl 2000: 294). Vienna's texture was thus located in an interplay between modernity/change and tradition/preservation, with no side winning or losing. Vienna was to stay in a permanent state of transition, or rather ambivalence – indeed, exactly of modernity – rendering the city a perfect laboratory for mapping changes in perceptions of urban space during the nineteenth century.

In 1868 the first *Vienna Baedeker* was published, providing the reader with a new way of experiencing and inhabiting the city:

> The more distances diminish, the more restless becomes the tourist . . . who today hardly has as many days left for a big city as he had weeks two centuries ago. In these circumstances the traveller is not served with a systematic (or even alphabetical) listing of sights only; he does not have the time any more to choose from the number of sights he is interested in and make routes of them. Rather, the travel book has to draw the routes within a big city for him . . . in similar fashion as long has been common for greater and smaller walks. . . . *Our 'Vienna'* makes the first attempt to solve this task.
>
> (Bucher and Weiss 1868, preface)[2]

I quote this passage from the *Vienna Baedeker* in detail, because it indicates a significant change in the way guidebooks have represented urban space to tourists. It clearly shifts the focus from a punctual perception of the city to a linear one. I have already shown (Spring forthcoming) how the act of moving from one sight to another was hardly if at all mentioned in early guidebooks and their enclosed maps from the late eighteenth until the mid-nineteenth century.[3] If it was, it was implied in a very different way, through the offering of exhaustive descriptions of sites (and sights) to see, and not through a conscious selection motivated by the tourist's time- and space-restraints.[4] By offering an encyclopaedic listing of sights, the city as a collection of image-dots, as pieces of information could only be linked together by the imagined space of the city of the map, and subsequently by the actual visit of the city. Early guidebooks thus tended to place objects in a more or less random series, leaving it up to the visitor to link them together in a spatial narrative. The links that were later to connect the objects – streets or medias of transportation such as the body or public transport – were missing or listed as pieces of information, not explicitly as modes of mobility, as was to be the case later. Eventually, in the course of the nineteenth century, guidebooks began to trace continuous lines and narratives through the city, either in the form of tours, or in the form of detailed information on how to get to the city itself and to one's accommodation, and to famous sights. In this new paradigm, guidebooks no longer only gave a means of reading the city (through a listing of objects) but of writing it by opening up the possibility of unnamed spaces lying between the individual objects on the tour. Links were established between sights, and the role of streets, squares and buildings connecting these sights changed in meaning and in significance.

Tracing Wider Lines: Tours through Europe

Of course, the history of thinking Europe as crossed and interlinked by tours reaches far beyond the nineteenth or eighteenth centuries. Pilgrimages are perhaps the earliest form of touring, with the traveller moving on a devised route from one place to another. The Grand Tour, popular since the sixteenth century, with its peak in the eighteenth century and its decline in the nineteenth century, followed (and reinscribed) a circuit of European cities, only slightly altering the stops on the route over the centuries. From the late sixteenth century onwards, travel handbooks provided the traveller with itineraries and descriptions of sights in urban space (Behringer 1999: 82–4; Towner 1996: 136), with Thomas Nugent's Grand Tour guidebook from 1756 suggesting various itineraries from town to town, and leaving it to the tourist to select them him- or herself (Towner 1996: 106–7). From the late eighteenth century onwards, tourism infrastructure developed rapidly, becoming increasingly institutionalized, with inclusive packages as its most prominent feature (see Towner 1996: 134–5). The number of tourists visiting Vienna, measured in the number of people staying in hotels, rose significantly from the eighteenth century onwards (see Towner 1996: 131): from around 146,000 in 1876 and about 307,000 in 1892 to almost 600,000 in 1910 (Kretschmer 1988: 76; Meißl 2000: 330).

However, not only the number of travellers, but also their expectations and perceptions changed in the course of the centuries. The tradition of collecting and systematizing objects, popular in the seventeenth and early eighteenth centuries, increasingly gave way to a search for the subjective content of travelling (Brenner 1999: 52–3). Travelling acquired a relational note: where travellers had before transferred the idea of their own culture onto other places, they were now interested in comparing their own cultures with other cultures, and this not only in spatial, but also in temporal terms (Brenner 1999: 51–5).[5] I would suggest that this relational perception of the surrounding world constituted one pre-condition for the sightseeing tour: only now could various objects in space be linked to one another so as to create such entities.

A Sense of Space Defined by Linear Movement

Researchers have argued against an overemphasis on new media of transportation, in particular the introduction of the railway, as an explanation for the changing structures of travel in the nineteenth century (Towner 1996: 110–14; Urry 1995). Indeed, while the development of the transportation system certainly played a relevant role in the choice of travel destinations, the particularity of places eventually proved to be more important than the comfort or speed of travel. A look at railway maps from the early nineteenth century informs us that the European railway system was still in its infancy in the 1840s, and that, hence, a smooth and continuous journey from one town to another was rarely possible (see Towner 1996: 114). Yet, as Schivelbusch shows, the introduction of railways led to a new perception of landscape and the surroundings in general. While the bourgeois traveller of the eighteenth century conceived of the places he or she visited as spatially individual and original, this changed with the fast movement of the train through the landscape:

> By destroying the part played by space and time in experience, the railways also put an end to the possibility of experience in the journey of education. From now on, places are no longer spatially individual or autonomous, but rather points of travel, contained by travel.[6]
>
> (Schivelbusch 2000: 173–4)

This shift to a relational and consequently also linear perspective may already be traced in the landscape paintings of William Gilpin and J.M.W. Turner, where place was seen 'as mediated by its connection to one place in the east, and another to the west'. This produces 'a sense of space which is defined always by this linear movement, so that to stop at a place is still to be in a state of potential motion' (Barrell 1972: 89, quoted in Aitchison *et al.* 2000: 38).

The new modes of transportation did not only alter the topography of towns in the mental landscape, but contributed to a significant change in the perception of the urban landscape and its sites as well. This shift can be traced in new urban traffic ticket systems, but also in the discussions of urban planners such as Camillo Sitte and Otto Wagner on the ideal city of the future and the association of traffic with modernity (see Meißl 2000: 291–302). Last but not least it may be seen in the new way guidebooks tackle the question of approaching and moving around the city. I have shown elsewhere (Spring forthcoming) that early guidebooks only mentioned in passing (if at all) the process of approaching the city and in particular of driving or walking across the city. Rather, a list of the means of transportation and the various places to find these means accompanied the guidebooks, leaving it to the traveller to combine this piece of information with the sights listed elsewhere in the guidebook.

This method of representation slowly changed in the course of the nineteenth century as the guidebook began to inform the traveller of the best and fastest way to get to Vienna, and how to reach certain places across and outside the city. The city could now be perceived as a net of lines (linear perspective) and dots (leading to a relational perspective), these becoming necessary orientation markers in the urban landscape. This shift in perception was first expressed in the wider context of inter-city travel, and slowly found its way into the intra-urban zone as well.

Until the late eighteenth century, urban or suburban traffic was of lesser significance than transnational traffic. This changed with the rapid growth of many European cities during the nineteenth century, introducing new questions as to how to structure circulation in the city. New streets were built, old ones were widened, and water and waste pipelines were erected with the aim of improving hygiene conditions. As a result of the extension of the economic market, rapid and unhindered transportation of commodities across urban space became a necessity. Already in the late eighteenth century an increasing number of the bourgeoisie could afford to travel to the country for pleasure, and, eventually, spending the summer months in the country became a status symbol for the new middle classes (see Czeike 1988: 44–52). From the early nineteenth century onwards, the so-called lower classes also discovered the joy of going for a day to the country and enjoying the fresh air, though it would take almost a century before public transport became affordable for the majority of the population. This change in leisure patterns was reflected in the growth of organized transportation. In 1834, public carriages (Gesellschaftswägen) drove to not less than 88 places outside Vienna on a regular basis (Czeike 1988: 58). In 1870, the guidebook even recommended reserving places on the carriages in advance, if the traveller intended to visit a popular place on a Sunday or a holiday (Bucher and Weiss 1870: 188). Thus, the tourist was forced to plan well ahead and to imagine the trip to Vienna beforehand in detail.

During the nineteenth century, efforts increased to open up Vienna for traffic. Streets and railways were built across the city in order to link both the various districts with one another and Vienna with the world outside. Similarly, the system of driving had to be adjusted to the needs of the modern city and its demands for unhindered fluidity. In 1819, a decree was issued ordering traffic to drive on the right in Vienna and its immediate suburbs (Vorstädte) though not in its more distant suburbs (Vororte) and in the rest of the monarchy, where one was to drive on the left. Most likely as a result of the increase of traffic this system had to be abandoned, and in 1852 driving on the left side was introduced in Vienna as well (Czeike 1989: 12–13). In 1880, laws were passed with the aim of keeping the traffic running fluidly. Not stops or traffic jams, but smooth lines were asked for, and as a result, certain carriages (Stellwagen) had to drive at a trot to keep the traffic moving (Czeike 1989: 18).

The public tramway system did get a boost from 1897 onwards with its communalization and beginning electrification, and its extension required a new mode of coordination that demanded of the passengers a certain flexibility and adaptability (Meißl 2000: 336–8). From 1903 to 1913, 325 million journeys were made using Vienna's public transport system (Czeike 1983: 7).

Visible signs of the change in perception can be detected in the newly introduced public transport lines using tramways and omnibuses and, later, trains. Unlike individual urban traffic such as the fiaker (hackney carriages) and sedan-chairs, these had signs on them informing the traveller of the first and last stop of the line and, often, of the stops in-between. The traveller was to follow a pre-given line through the city, exactly as he or she was used to from cross-country transportation systems such as stagecoaches and trains. While booklets informing the traveller about the distances between various towns existed already in the sixteenth century, and the post informed the traveller of itineraries and the stops along the route since the early seventeenth century

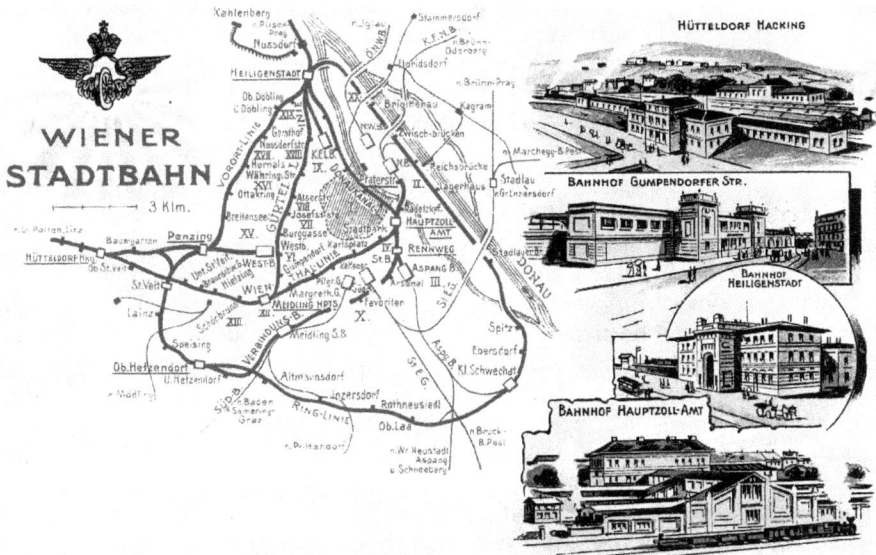

Figure 2.1 Map of Vienna's City Train, Postcard, c.1900, Wien Museum

(Behringer 1999: 80–6), for a long time urban public transport did not have fixed stops for getting on and off, but stopped according to the desires of the traveller. In Vienna this changed slowly from the early nineteenth century onwards, when carriages to the surroundings of the city introduced fixed stops along the line and followed a fixed schedule. From the mid-nineteenth century this system also was applied to traffic across the city (Czeike 1987/8: 4–5).

The traveller (and the local) thus had to think in advance of the places they were to reach and to decide where to get off. Indeed, they had to internalize the transportation system and the urban network to a certain extent in order to get to the right places (see also Spring forthcoming). The city turned into a space marked by dots and connected by lines. This linear and relational way of perceiving urban space was further implemented in Vienna through the introduction, in 1903, of tickets that could be used to change from one line to another (Czeike 1983: 8). These developments in transport policy reflect tendencies in urban planning making the passenger's or driver's transfers across the city as smooth and as quick as possible. Otto Wagner's model of Vienna had as its basis an indefinite net, crossed by the circular and radial lines of the urban railway, making it possible to reach any dot on the urban map by changing train just once (Meißl 2000: 301).

The 'Representation Crisis' of the City

In the mid-nineteenth century Vienna consisted of three major areas: the old town, the suburbs within the embankment (Vorstädte) and the outer suburbs (Vororte). In 1850, the first extension of Vienna took place, the second following in 1890–2, and the third in 1904–5. In 1890, the outer suburbs became part of the City of Vienna and, as a result, the population rose from around 800,000 to approximately 1.3 million. With the incorporation of the suburbs, the demolishing of the fortifications and the forcing of streets and traffic across these three zones, Vienna's space became more difficult to organize and to deal with. The fragmentary character of the expanding city and the necessity of imagining the modern city in a different way have often been commented on (e.g. Boyer 1996; Donald 1999: 3–4). As a result of the growth of the city, it was no longer possible to imagine the whole city at one glance; rather, one had to take parts of it and construct the entity of the city from different experiences. In other words, the city had turned into a collection of metonyms. For the tourist this meant being confronted by a maze of streets, roads and houses, all making it difficult to get an overview of the city. My argument is that tours helped structure this experience of complexity, offering a tool for getting through the city and collecting sights on the way while retaining orientation and control. A precondition of all these activities, however, was that the streets be in good enough condition. Moreover, streets and squares were to become important signs of orientation and thus aids in structuring the mazes that expanding cities became in the late nineteenth century.

Streets were widened during the nineteenth century, increasingly paved and adjusted to the needs of the pedestrian and of traffic in general. The city needed the newly acquired space for accommodating the rising demands of traffic, but urban planning also had as its aim the exertion of social control and improvement of hygienic conditions. Not surprisingly, the projects for the extension of Vienna submitted in 1858 overwhelmingly focused on the role of street-traffic routes in their schemes (Breitling 1980: 36).

Accordingly, while in the nineteenth century guidebooks tended to define Viennese streets primarily with regard to their bad state or narrowness and their role as means of reaching certain places (e.g. *Neuester Wienerischer Wegweiser* 1797: 5–6; Weidmann 1859: 38–40), streets were slowly assigned a different role in guidebooks at the turn of the twentieth century. *A. Hartlebens Illustrierter Fremdenführer* (Illustrated Visitor's Guide) from the early twentieth century illustrates vividly the rising significance of the street as a medium of orientation and as an aim in itself within urban perception, with its role as link between certain places and as medium of movement strongly emphasized. Just as a traveller in the present, Michel de Certeau, climbed to the top of the World Trade Center and looked down on the city, turning it into the text 'New York', tourists to Vienna in the early twentieth century were to climb to the top of the highest tower in town and to transform the city into a readable text and the text of the guidebook into the city. The bird's eye view of the city, in de Certeau's words, 'makes the complexity of the city readable, and immobilizes its opaque mobility in a transparent text' (1988: 92). The guidebook thus asked the tourist to look specifically for streets from the top of St Stephen's Cathedral, giving directions based on their coordinates, and in a chapter called 'Walks to look at streets and squares' it sent the tourists on a tour through the city by asking them literally to follow streets and to turn into other streets at certain numbered houses: 'From Stephansplatze at nr. 7 into Rotenturmstraße, from there at nr. 15 to Lichtensteg and Hohen Markt, at nr. 5 into Tuchlauben' (*A. Hartlebens Illustrierter Kronenführer* [1907/8]: 27).[7] Only after the tourist had acquired a feeling for the city in this way did the guidebook recommend tours that actually aimed at sightseeing in its conventional sense.

But how does this abstract view of streets translate into the practice of actually walking the streets? After all, the tourist standing on top of the tower of St Stephen's Cathedral, looking down on the net of streets spreading out below him or her, waiting to fill the geometrical space with sights and individual experiences that he or she later is to acquire on the tours recommended by the guidebook, appears to control the city with his or her gaze, submitting its chaotic life to the systematic grid of the map and the guidebook's depiction of ordered lines.

De Certeau's concept of enunciation can help explain this transferral, it being a means for translating the various representations of urban space as we meet them in the guidebook – maps, written texts, images – into the actual experience of the city. To de Certeau, because 'space is a practiced place', 'an act of reading is the space produced by the practice of a particular place: a written text, i.e. a place constituted by a system of signs'. In the same vein, 'the street geometrically defined by urban planning is transformed into a space by walkers' (1988: 117; see also 97–8).[8] Thus, the reading of literature always clears the way for a set of cultural practices (de Certeau 1988: preface; see also Koshar 2000: 5–6). De Certeau opens a gap where the immobility of the written words of the guidebooks and the static lines and curves of the map can fold into the traveller's experience of the city, its buildings, its people, its atmosphere.

Thus I would suggest that the city becomes a rich texture to be read in a satisfactory way by the tourist only when the various spatial readings of the city are juxtaposed and combined, allowing them to intersect. The travel guidebook offers one way of gaining access to this polyvalent reading of the city. Furthermore, by applying guidebooks to the city, the tourist not only reads but also writes (re/produces) its spaces. *A. Hartlebens Illustrierter Kronenführer* provides an example of this translation from reading place into performing space with the following advice to the traveller who had just climbed the highest tower of Vienna, i.e. the top of St Stephen's Cathedral: 'With the help of a map and explanations gladly given by the tower watchman in return for a small tip, it is not difficult to acquire a general knowledge of the structure of Vienna, to which the details acquired on the walking tours may be added' (*A. Hartlebens Illustrierter Kronenführer* [1907/8]: 24–5).[9] The traveller thus lives the city by looking at it and again by fitting the structures just observed onto the busy life of the streets. De Certeau has been criticized for producing a binary opposition between the view from the tower and walking in the streets (for an overview of this criticism see Donald 1999: 14). However, as the view from the tower recommended in the guidebook shows, this binary opposition is dissolved in the tourist's practical experience of the city. This is because the tourists were to look at the city from above, fixing it as a mental map in their imagination, and then to transfer the lines of this map onto the real space of the city in order to get to know and to understand the city's structure. Indeed, eventually the tourist was to inscribe the space of the city with his or her steps (and texts). I want to suggest here that tours functioned as a means to ease this transferral, or indeed, to make it possible.

Compare this with another view from a Viennese tower recommended 100 years earlier: in 1803 Anton Reichsritter von Geusau published a guide to Vienna, basing his 'Descriptions of the Panorama of Vienna' on a map enclosed in the guidebook. This map by the Englishman William Barton showed a panorama of Vienna based on the view from the Church of St Augustine, taken by von Geusau as a starting point for his descriptions (von Geusau 1803: 24):

One sees in the foreground the city of Vienna with its beautiful towers, palaces, houses, squares and streets; bridges, the Esplanade. The suburbs lying in a circle around the city; the stream Danube rich with shipping; then the region around Vienna with its magnificent summer castles and residences . . . until finally the far away mountain ranges from the

Schneemountain to the Haimburger- and Preßburger mountains finish
this remarkable circular painting.[10]

(von Geusau 1803: 24)

Von Geusau's aim is to look across the whole city 'from the left to the right'
and to explain 'the most remarkable objects' illustrated on the map (von Geusau
1803: 25). In other words, the reader is invited to look at the map and to imagine
its buildings and drawings as real while reading the accompanying descriptions
of the various buildings. The circularity of the map stands in marked contrast
to the linearity of the narrative put forward by von Geusau. However, the main
difference to later tours lies in the detailed description of what is to be seen
on this map, and in the negligence concerning how this description actually
would translate into the practice of the visit. I would like to suggest that one
explanation lies in von Geusau's predominantly punctual way of perceiving the
city, i.e. where the buildings are the main attraction and the links in-between
insignificant, nothing but a necessary means of reaching them. In contrast,
streets in the late nineteenth century were a symbol and metonym of movement,
as the tourist was not only expected to look at the network of streets from a
bird's view or via representations of space such as maps, but also to walk along
them in order to get a feeling for the city and to reach specific sights.

The city turns into a text that is to be read and to be written on; the spatial
practice of walking the city folds into the written text of the guidebook. This
indicates that the process of translation is facilitated by the mode of perception of the city, that is, by perceiving it as a grid of streets and roads that the
tourist has to pace along in order to get control of the complexity of the city.
Walking, and indeed, all movement through urban space thus becomes a crucial
aspect of the process of consuming the city, and of producing its infrastructures
of intelligibility such as streets, signs, numbers and names.

As the quote of Hartleben's guidebook shows, not only streets but also street
names and house numbers were an important means of orientation in the growing city and of controlling the urban complexity. While the idea of numbering
houses emerged in Vienna in the sixteenth century, and the first book to describe
houses along streets dates from 1701, a consequent numbering of houses did
not take effect before 1770 with a patent by Maria Theresia (Csendes and Mayer
1987: 4–5). Significantly, the aim behind this first consequent numbering of
houses was to list people for conscription, i.e. to find a means of categorizing
them. Accordingly, the numbers were named 'Konskriptionsnummern'. A century later, in 1860, the aim of numbering houses had changed to that of orientation, the numbers now being called 'Orientierungsnummern'. The desire for
orientation in the growing and rapidly changing city of the eighteenth century
was also articulated in another context. From 1782 onwards street names had to
be painted on the walls, and in the following decades street plates were put on
each end of the street, functioning as signs of orientation particularly on squares,
i.e. centres of a network of streets (Csendes and Mayer 1987: 5). With the expansion of the city from the mid-nineteenth century onwards, it became necessary

to introduce a new system of numbering that followed the way houses were lined up along the street: in 1862, houses in Vienna were numbered along the street they were situated in, radiating and diagonal streets were differentiated by being assigned differently shaped street plates, and the various districts received differently coloured street plates – the city literally became visually decipherable. In 1872, streets with the same names had to be renamed and in 1894, streets (Gasse) and roads (Straße) were distinguished for the first time, the roads having a width of at least 16 metres (Csendes and Mayer 1987: 8–9). Hence, street names as well as house numbers increasingly became, both for the tourist and the local, a means of orientation in the city – rather like the signposts of today's tourist areas – and were assigned a prominent place in the introductory chapters of the guidebooks (e.g. *Kaiser-Jubiläums-Führer* 1898: 10). In 1927, the Vienna magistrate rejected the proposal by the *Deutscher Schriftverein Wien* (The German Society of Letters in Vienna) to reintroduce the Gothic Fractur script on street plates because of the difficulty visitors from abroad would have had with deciphering it (Csendes and Mayer 1987: 11).

Street names did not only provide a means of orientation in space but also in time. In 1894, the magistrate decided to name big squares and important streets exclusively after famous persons and by 1867 the Viennese *Gemeinderat* (the town council) had arranged for plaques to be put on houses (Bucher and Weiss 1870: 65), in order to commemorate certain people or events. The traveller arriving at these places was thus given a narrative that linked the city with a certain culture and history, reinforced by the significance of respective squares and streets. One may deduce from this increased number of urban history identification tags a growing number of tourists exploring the city on their own – and on foot.

Walking in the City

Vienna Baedeker from 1868 explicitly based its city tours on tours on foot (*Fusspartien*), calling them *Wanderungen*, a term that in German denotes walking in the country, i.e. rambles or hikes. Here, *Baedeker* was following a tradition that had characterized travel at least since the onset of romanticism in the 1820s and 1830s, when walking tours in Switzerland became part of the Grand Tour (Towner 1996: 114–15). Rousseau had opened the path for a new form of travelling: 'Taking the place of utilitarian orientation, which presupposed the privileging of the destination before that of the way taken, we find a meandering rambling which seeks pleasure in the self by seeking pleasure in nature'[11] (Brenner 1999: 57). In other words: no longer was the aim of the journey important, but the way itself.

Walking in the country, as it was practised from the late eighteenth century onwards, has often been depicted as a means of contesting or resisting existing social norms (Jarvis 1997). Already in the early nineteenth century the freedom it conveyed was diametrically opposed to the packaging of tourists onto pre-determined tours (Lewis 2001: 64–5), and, we may add, to the rushing of tourists along devised city paths. Thus, while in this context, walking is considered mainly as freely wandering around, with the walker in the country following a non-linear trail (Jarvis 1997: 56), the walking tourist tracing the city tours prescribed by guidebooks entered a path of linearity and circularity, leading him or her to particular places that stringed together the various images of the city.

A more important predecessor of urban sightseeing tours appears to have been excursive walking, i.e. walking along paths that had been registered and recommended, a practice that rose in popularity during the nineteenth century (Wallace 1993), reinscribing paths of continuity and recognition into the landscape. Urban sightseeing tours functioned in a similar way, as they too repeated and familiarized the strange and unknown landscape of the city. However, I would like to suggest that urban sightseeing tours combined these two modes of walking: the solitary mode, which offered the freedom of walking around at one's own pace and making detours from the recommended path, and the collective mode, where the tour's path functioned as a beaten track, a path already worn out by the number of its readers.

Although guidebooks for travelling on foot had their peak in popularity at the end of the nineteenth century, they were already being published in the early years of the century (Koshar 2000: 34–5), with the reader being able to choose among various tours. A booklet by Franz de Paula Gaheis, *Wanderungen und Spazierfahrten in die Gegenden um Wien* (Rambles and Walking Trips in Vienna's Surroundings, 1797–1808), had its fourth edition as early as 1809, and was a major force in the popularization of leisure walking in the vicinity of Vienna (see Czeike 1988: 48). In 1839, F. C. Weidmann published his *Wien's Umgebungen* (Vienna's Surroundings) introducing ten different tours for discovering Vienna's immediate surroundings, and Schmidl's *Wien's Umgebungen* (Vienna's Surroundings) from 1838 took Vienna as starting point for several excursions. Walking became an integral part of urban experience from the 1840s onwards – people flocked to the countryside for a Sunday trip, and summer houses were built on the edge of the city.

Walking in the close vicinity of the city had been popular in Vienna from at least the late eighteenth century, when Joseph II had opened several parks and places (such as the famous Prater in 1766 and the Augarten in 1775) to the public. Promenading on the ramparts (Bastei), which until 1817 functioned as part of the military defence for Vienna, was already a favourite pastime of many Viennese in the eighteenth century (*Neuester Wienerischer Wegweiser* 1797: 6; see also Birkner 1998: 30), and when from 1858 onwards the ramparts were being demolished, many Viennese lamented the loss of their favourite place of walking and relaxation (Birkner 1998: 30–1; Breitling 1980: 39).

Walking for pleasure in the city however was a phenomenon of the late nineteenth century. As shown earlier, walking in and across the city has as its precondition a certain space assigned to pedestrians. Vienna did not traditionally possess a great number of large squares, and complaints about the traffic system, in particular about the recklessness of many drivers, were to be found throughout the eighteenth and nineteenth centuries (see Czeike 1989). As a result of efforts by the government, urban space was increasingly opened up for pedestrians from the 1860s onwards, and in 1911 projects were introduced aiming at providing pedestrian crossings in the city (Czeike 1989: 6). Another aspect that may have prevented earlier leisurely walks through the city was the rise of anti-urban tendencies parallel with the increased interest in the countryside in the Romantic period (see Towner 1996: 121), providing the basis for a commonly held association of walking in the country (or climbing up mountains) with the cleansing of both body and mind, in contrast to the unhealthy act of walking in the city (see Edensor 2001: 85; Lewis 2001: 59).

In fact, although walking through the city certainly had been a part of the visitor's itinerary for a long time, it is only from the 1860s onwards that walking became an explicitly and frequently stated part of visits to Vienna. Hence, I want to suggest that along with the opposition of country versus city, one may observe a process of transferral from walking in the country to walking in the city. The guidebook in 1868 sent the tourist on tours of a type devised for rural discovery and conquest, now transformed and adapted to the landscapes of the city. In other words, we may speak of a transferral of an already existing rural concept onto the urban landscape.[12] This recalls the quote by *Vienna Baedeker* popularizing the concept of urban tours to the tourists: 'Rather, the travel book has to draw for him the routes within a big city, and in a similar fashion as it long has been common for greater and smaller walks.'

As mentioned earlier, the experience of walking in the city may turn into an act of reading and writing the city, of observing or partaking in the everyday life and even of being-in-the-world. Walking in the city may create a performative or enunciative space that integrates a decisive element of subversity (de Certeau 1988), or be a means of both contesting and confirming existing regulations (Edensor 2000; see also Fyfe 1998). Certainly, guided tours usually do not open for contestations of urban spaces, but rather for their affirmation. This is because the guidebook foresees potential deviation from unidirectionality. By focusing in its descriptions of the city on directing signs, such as street names, street numbers or certain sites interpreted as sights, they make sure that the tourist cannot go astray. Guided tours, by leading the tourist from one sight to another on a well laid-out path, only affirm the abstraction of the tourist from busy everyday life in the streets. The mobility of the tourist appears to be restricted to a laid-out path, but the path also makes it fluid and unhindered by interdictions, byways and other obstacles on his or her way. In that way, tours create a particular tourist space, marked by rules, clarity and predictability, which differs strongly from other spaces (such as that of the everyday) where choice, arbitrariness and interruptions may prevail. Looking at guided tours from this perspective, the world for which the tourist following the devised path is looking is systematized in advance, 'prepared in consumable space-time units'[13] (Achleitner 1996: 201–2). The subversive character of walking is diminished in so far as the directions are pre-given and the objects to look at are written out beforehand. The tourist is asked to look ahead, to reduce urban space to a line leading him or her from one object to another.

However, as discussed earlier, walking through the city involves always a process of mediation between the urban fabric and subjective experiences arising through stories and histories. It is at this intersection that the process of imagining the city is situated, interrupting the unilinearity of perception just sketched out. As much as tours draw pre-given lines and curves through the city, they also open up for detours and places that may not have been part of the tourist itinerary. In fact, the *Vienna Baedeker* encourages the tourists to create their own (tourist) space, thus writing the city according to their own ideas and wishes: the proposed tours should 'of course not inhibit anyone to make a rest, to give up some tours, or to extend others'[14] (Bucher and Weiss 1868, preface). Guidebooks thus did not deterministically exert control over their reader. However, at the same time the very technology of the guidebook imprints the tourist with an internalized imagination of the city, of the routes through it, and of the ways of making it intelligible. While the tourist may depart from the suggested route, he or she cannot really depart from the ways of thinking and narrating the city that the guidebook implements. Indeed, these forms are inscribed on the material city itself through the infrastructures of pavements, signs, street numbers, maps, etc.

Moving across Public Space

The nineteenth century is renowned for its tendency to apply categories to people and to construct schemes that act normatively rather than descriptively, as Foucault so aptly has shown. As a result it opened the way for the exercising of control, but also the extension of social welfare. The concept of 'tourist' is just such a category, moving from the rather elusive definition of the traveller to the more exclusive category of the tourist. Tourist spaces, such as hotels or spas, were designed to accommodate the tourist's needs and desires, to provide the tourist with a space where he or she may move in relative security, re-inscribing the lines drawn for him or her by other tourists or at least by the writer of the guidebook himself. Of course, often this space overlapped, merged or had blurred borders with other spaces. Nevertheless, reading urban space from the perspective of the tourist creates different readings than when it is read from the perspective of other categories of people.

As we have seen, Vienna underwent dramatic growth during the nineteenth and early twentieth centuries, both in terms of area and population. From the late eighteenth and early nineteenth centuries onwards, urban public space was redefined in a profound way. Bourgeois spaces, public parks, cafés and shops, emerged all around the city and became essential environments for leisurely walks. At the end of the nineteenth century, public space had been extended to also include 'people's space'. With the rise of the welfare state, with the spread of new technologies of communication and transportation, sites of modernity such as railway stations, public gardens, bridges, etc. were recast or built from scratch as overt statements of the state's care for its citizens (Boyer 1996). The development of public transport also created an extended public meeting-space, as an increasing amount of the urban population spent time in carriages together with unknown people. This followed directly on the tradition of the postal network as a means of democratization and egalitarianism in the seventeenth and, particularly, the eighteenth century, when women, children and sick people were allowed to travel by post carriages (Behringer 1999: 88, 93–4). A liminal space came into being which blurred the border between public and private space, transforming urban exterior space – streets and squares – into interior spaces (Holston 1989; Leslie 2002: 65). This rise of a liminal city space may have contributed to the introduction of tours into the urban landscape, as the city lost – at least in certain areas – some of its dangers and threats, turning into a space in which an increasingly growing number of civilians could participate. Also, with the increased ordering and structuring of the city as a result of urban planning, the ideal environment for the tourist had been created, and the tours affirmed this by showing in clarity the safe way through the ordered maze of urban space.

Public space – the space the tourist is allowed to move in – is defined by the buildings that border it, by the stories that are told about it, and by the tours that trace their paths across it. Here, the tourist's social world is produced and reproduced. Yet these spaces are often ascribed specific meanings that are crucial for the collective memory and perceptions of the inhabitants as well (see Spellman 1996: 148). As Meißl (2000: 364) observes for the period around 1900:

> The inhabitants of Vienna found themselves increasingly confronted with a modular metropolitan urban framework, where – relative to the earlier localistic narrowness and coherence of life – spaces of work, living, leisure time, consumption, etc., and also the spaces of self-movement or of being moved between these spaces, replaced one another with an often confusing complexity and speed according to continually changing temporal rationalities.[15]

The various spaces of the city expanded and changed shape and character faster than before. Crossing over from one space into another became easier and thus an increasing number of places were opened up to the tourist and the local alike.

Hence, specific local spaces may turn into generally accessible spaces where the tourist learns to understand the city, to infiltrate its silent places, inscribe them into the transparent pages of tourist memory. The tourist moves from one public space to another, from one major square to a promenade, in order to acquire a feeling of the civic life of the city. This suggests that the growth of 'civic spaces' (Boyer 1996: 8) was a pre-condition for opening up of urban space so that tourists could trace their routes through the city. Accordingly, the tourist's public space turns into a place of intersection and intermediation between the foreign and the local. Guidebooks took up this polyvalent meaning of space by addressing themselves frequently to both the foreign visitor and the local population, in their titles or forewords (e.g. *Neuester Wienerischer Wegweiser* 1797).

Accelerating the Tourist Experience: Time

As mentioned earlier, tours leading from one sight to another existed in the city long before they entered the guidebooks. Fiaker (drivers of Fiaker, i.e. hackney carriages) were popular guides in Vienna, and professional guides offered their services to the tourists. Guides were a common feature of Italian cities already in the seventeenth century (Towner 1996: 136). However, guidebooks introduced a new dimension in the history of guiding, as one explicit aspect of the guidebook was that it should help the tourist to find his or her way through both the countryside and the city without having to ask locals for information or having to pay for a professional guide (see, e.g. Schmidl 1833: 319). Schmidl's guidebook observes that as a result of the development of paths across Viennese countryside, the necessity to engage a human guide, still necessary in the 1820s, disappeared in the 1830s (see Czeike 1988: 62). Also, as a consequence of the development of city infrastructure and internal routes, together with the spread of the guidebook, the demand for city guides decreased dramatically, and the costs of the journey decreased accordingly (see Koshar 2000: 30–1). At the same time, the length of time the tourist was to stay in a city changed as well, gradually diminishing from several years in the high period of the Grand Tour to several weeks, several days or even a single day in the later part of the nineteenth century (see Towner 1996: 132). As a result of the expansion of the city and the reduced time-frame of the traveller, visiting the city became a more time-consuming and complex process. Clearly, the introduction of guided tours was the guidebook's response to the tourist's growing demand for rapid orientation in an unknown space.

My material suggests that this compression of time applied to all kinds of guided tours, whether on foot or by cab. I therefore do not fully agree with the argument that walking tours generally are signs of the tourist's desire for slowness, for finding a space that counteracts the hurried pace of tours by bus or tram (Koshar 2000). As the quote by *Vienna Baedeker* indicates, walking tours were established in order to meet the tourist's desire for efficiency, for being able to rush across the city within a few hours and to gather as many impressions as possible. Thus, they are a sign of the growing emphasis on mobility that characterizes the nineteenth century.

Yet, one should also be aware that the tours offered in the guidebooks often did not demarcate the line between walking and other means of transportation. Indeed, both walking and other forms of transport were often recommended as part of one single tour, thus combining very different experiences of urban space in terms of: (1) movement, i.e. moving the body vs the body being moved (see

Edensor 1998; Larsen 2004); and (2) perception, i.e. different ways of gazing or glancing at the surroundings (see Urry 2002; Larsen 2004). A definite change in temporal flows became obvious only in the late nineteenth century, when tours on foot could be juxtaposed with those by tram or car. By the early twentieth century private agencies and the public transportation system took up the idea of offering guided tours by tram or by car to the tourist. Guided tours were collectivized, institutionalized and made into an integral part of the touristic experience (*A. Hartlebens Illustrierter Kronenführer* [1907/8]: 30; Busson 1913). This was partly an outcome of the increasing differentiation of tourists around 1900, as statements by guidebook author Paul Busson (1913: 3) illustrate. He distinguishes between the 'rushing globetrotter who sacrifices only a few days to the city, where his joy of travelling leads him'[16] and the 'real traveller who comes here to really get to know the famous imperial city and to reach into its innermost being'.[17] Guidebooks increasingly recommended walking tours mainly to those tourists who had sufficient time to spare. At this stage, of course, Koshar (2000: 34–5) is right to emphasize the character of walking tours in the city as a mode of contemplation, as a response to the ever-growing velocity of travel.

Linking the City

As described above, guidebooks from the late eighteenth and early nineteenth centuries would provide the tourist with a list of sights, spreading them like dots in the urban landscape. In the process they paid little, if any, attention to the links between the dots, that is, to the tactics the traveller had to employ in order to get from one place to another. Even though some guidebooks provided the reader with arrival and departure details already in the early nineteenth century (Boyer 1996), information on local and state traffic in the form of an extensive and detailed chapter on the matter did not enter guidebooks before the mid-century. My argument is that this shift has to be viewed in the context of social and cultural changes associated with modernity (see also Spring forthcoming). In modernity, the demand for progress and search for national destinies was equalled by the desire to build representative and representable capital cities. Urban planning, and in particular the development of various forms of circulation (sewerage, boulevards, communication technologies, traffic) became a means of implementing health policies and of anticipating rioting, particularly from the working class, in increasingly crowded and steadily growing cities (see Boyer 1996). Just as Haussmann cut Paris into boulevards, from 1858 onwards Vienna's city wall had to give way to the Ringstraße with its imposing new buildings. While the re-arrangement of Parisian urban space allegedly led to the death of the flâneur, it can be said that the building and widening of some Viennese streets opened up spaces for the tourist to walk in. The Ringstraße became Vienna's boulevard, flanked by Palais (town houses) and public buildings that invited the (wealthy) pedestrian to pause and admire.

The sights increased in variety during the nineteenth century, opening up new spaces for the tourist, and widening his or her perspective of the city. In parallel with the increase in available space, the amount of time for a visit decreased. It became necessary to order sights according to conventions of interest and relevance in order to accommodate the early mass tourist's time-frame and various desires. No longer were the various sights of the city distributed randomly across urban space; rather, they were transformed into narrative sequences. These urban spatial narratives often appear to have consisted of rather arbitrary collections of objects. A tour recommended by the *Kaiser-Jubiläums-Führer* from 1898 illustrates this arbitrariness: it traced lines and circles through urban space, leading the traveller from St Stephen's Cathedral to the department store Philipp Haas & Söhne, to a monument dedicated to Mozart, to the Imperial palace and finally to a promenade along the recently built Ringstraße (1858–65) with its imposing new buildings (*Kaiser-Jubiläums-Führer* 1898: 16–44).

Today, tours cut through all of Europe, in the form of cultural heritage trails or European walking tours; they are a feature of villages that want to have a claim on being of tourist interest, and their number is steadily increasing in cities. Just take a look at a webpage of any European city, and you will see maps with lines tracing their way across streets and past buildings (for Vienna see the entry 'discovery tours' at www.wien.info). While the tours I discussed in this essay are connected with a specific juncture of modernity, the survival of the form may partly involve refunctionalization into later modernity contexts, but also signify the 'anachronistic' or 'intertextual' survival of older forms of modernity alongside late modernity. Crang's (1996: 441) perception of a 'geography of the city structured around tours and heritage walks that string together the "historic" places of the city through their spatial arrangement rather than through thematic connections' aptly describes the tours of 1900, and also those tours in the present that aim to show the city in a nutshell (bus tours in particular) – though it does not apply to such an extent to the majority of tours taking the tourist on thematic trips into Baroque, Red and Jewish Vienna, or on the ghost hunts and underground explorations that have recently become so popular in many European cities. Tours reproduce the various representational forms used to picture the city, and at the same time open up the possibility of actively enacting or writing the urban space into which they are embedded. Their gaze directed ahead, tourists are to conceive of space in terms of discrete yet linearly ordered and connected objects creating the spatial form of the city.

Acknowledgements

I would like to thank the editors, Sándor Békési, Johan Schimanski and the participants of the 'Alternative Mobilities Future' Conference at Lancaster University for helpful comments on earlier versions of this chapter.

Notes

1 'was für Ausländer, besonders aber für Amerikaner ... die größte Anziehungskraft besitzt, und zwar die alten Stadtviertel Wiens mit ihren engen, winkeligen Gassen.' All translations by the author.

2 Je mehr die Entfernungen schwinden, desto ruheloser wird der Tourist, ..., der heute für eine grosse Stadt kaum so viel Tage übrig hat, als Wochen vor zwei Jahrzehnten. Unter diesen Verhältnissen ist dem Reisenden mit einer systematischen (oder gar einer alphabetischen) Aufzählung der Sehenswürdigkeiten allein nicht mehr gedient; er hat nicht mehr die Zeit, aus der Menge das auszuwählen, was ihn interessirt, und daraus wieder Routen zusammenzustellen. Das Reisebuch muss ihm vielmehr die Routen innerhalb einer Grossstadt vorzeichnen, ..., in ähnlicher Weise, wie das für grössere und kleinere Fusspartien längst gebräuchlich ist. Unser '*Wien*' macht den ersten Versuch, diese Aufgabe zu lösen.

3 The definition of 'tourist guidebook' is a rather elusive one as it may comprise all books aimed at guiding the traveller, and in this case they emerged as early as the sixteenth century. However, in this context I am interested in guidebooks as a genre that was shaped in the eighteenth century and – more or less – has survived almost unchanged right into the present.

4 But see Schmidl (1840: 353–62) for an early example of organizing space around tours.

5 Brenner (1999: 52) even speaks of a change of paradigm in the second part of the eighteenth century, when travelling and self-education as a result of the encounter with other cultures became central.

6 Die Eisenbahn, die den Raum und die Zeit für die Erfahrung vernichtet, beendet damit auch die Erfahrungsmöglichkeit der Bildungsreise. Von nun an sind die Orte nicht mehr räumlich individuell oder autonom, sondern Momente des Verkehrs, der sie erschließt.

7 'Vom *Stephansplatze* aus bei Nr. 7 in die *Rotenturmstraße*, von dieser bei Nr. 15 auf den *Lichtensteg* und *Hohen Markt*, bei Nr. 5 in die *Tuchlauben*.'

8 Since de Certeau's writings are in French, some of his terms slightly change their meaning when translated into other languages, such as the term *espace*, which may translate as both *place* and *space*.

9 Mit Zuhilfenahme eines Planes und den gegen ein kleines Trinkgeld bereitwillig gegebenen Erläuterungen des Turmwärters ist es nicht schwierig, sich eine allgemeine Kenntnis über die Struktur von Wien zu erwerben, an welche sich dann die bei Spaziergängen erworbenen Details anreihen.

10 Man sieht im Vordergrunde die *Stadt Wien* mit ihren schönen Thürmen, Pallästen, Häusern, Plätzen und Straßen; Brücken, die *Esplanade*. Die im Zirkel um die Stadt liegenden Vorstädte; den schiffreichen Donaustrom; dann die Gegend um Wien mit ihren herrlichen Lustschlössern, Landhäusern ... bis endlich die weit fernen Gebirge vom Schnee- bis zum Haimburger- und Preßburgerberge das merkwürdige Rundgemähde schließen.

11 'An die Stelle der Zweckorientierung, die eine Unterordnung des Wegs unter das Ziels (!) voraussetzt, tritt ein mäandrierendes Wandern, das den Selbstgenuß im Naturgenuß sucht.'

12 We can also trace this transferral from tours in rural landscape to those in urban landscape in a shift in terminology: the word *tourist* applied for the greater part of the nineteenth century to people walking in the mountains. Only slowly was it extended to also include visitors of cities. The contemporary German word '*Stadtwanderung*' for 'city walk' still bears witness to this adaptation, since, as mentioned above, '*Wanderung*' by itself is associated with walking in nature.

13 '... in konsumierbaren Raum-Zeit-Einheiten zurechtgemacht.'
14 'natürlich Niemand (hindern), Ruhepausen eintreten zu lassen, einzelne Touren aufzugeben, oder andere auszudehnen ....'

15   Die Bewohner Wiens sahen sich zusehends mit einem modularen metropolitanen Stadtgefüge konfrontiert, wo im Vergleich zur früheren lokalistischen Enge und Kohärenz des Lebens Räume des Arbeitens, des Wohnens, der Freizeit, des Konsums etc., Räume des Sich-Bewegens/Bewegtwerdens zwischen diesen Räumen mit oft verwirrender Komplexität und Geschwindigkeit mit immer wieder wechselnden Zeit-Logiken einander ablösten.

16 'hastige(n) Globetrotter, der jeder Stadt, in die ihn seine Reiselust führt, nur wenige Tage opfert.'
17 'eigentliche(n) Reisende(n), der hierher kommt, um die altberühmte Kaiserstadt wirklich kennen zu lernen und in ihr inneres Wesen einzudringen.'

Works Cited

A. Hartlebens Illustrierter Kronenführer: Wien, Illustrierter Wegweiser durch Wien und Umgebungen. Mit 60 Illustrationen, zwei Plänen im Texte, einem Plan von Wien und einer Karte des Semmerings (1907/8) 10th edn, Wien, Leipzig: A. Hartleben's Verlag.
Achleitner, F (1996) 'Lebensraum und/oder Ausstellung. Zur Wahrnehmung altstädtischer Räume. Das touristische Paradoxon', in *Wiener Architektur. Zwischen Typologischem Fatalismus und Semantischem Schlamassel*, Böhlau, Wien: F Achleitner, pp. 201–10.
Aitchison, C, MacLeod, N E, and Shaw, S J (2000) *Leisure and Tourism Landscapes: Social and Cultural Geographies*, London and New York: Routledge.
Barrell, J (1972) *The Idea of Landscape and the Sense of Place, 1730–1840*, Cambridge: Cambridge University Press.
Behringer, W (1999) 'Reisen als Aspekt einer Kommunikationsgeschichte der Frühen Neuzeit', in M Maurer (ed.) *Neue Impulse der Reiseforschung*, Berlin: Akademie Verlag, pp. 65–95.
Birkner, O (1998) 'Im Schatten der Revolution. Zum Bau der Wiener Ringstraße', *Wiener Geschichtsblätter*, 53: 26–37.
Boyer, M C (1996) *The City of Collective Memory: Its Historical Imagery and Architectural Entertainments*, London, England and Cambridge, MA: The MIT Press.
Breitling, P (1980) 'The Role of the Competition in the Genesis of Urban Planning: Germany and Austria in the Nineteenth Century', in Anthony Sutcliffe (ed.) *The Rise of Modern Urban Planning*, New York: St Martin's Press, pp. 31–54.
Brenner, P J (1999) 'Der Mythos des Reisens. Idee und Wirklichkeit der europäischen Reisekultur in der Frühen Neuzeit', in M Maurer (ed.) *Neue Impulse der Reiseforschung*, Berlin: Akademie Verlag, pp. 13–61.
Bucher, B and Weiss, K (1868) *Wien. Führer für Fremde und Einheimische. I. Theil: Wiener Baedeker. Wanderungen durch Wien und Umgebungen. Mit 2 Stadtplänen, 6 Theaterplänen und 30 Holzschnitten*, Wien: Tendler & Comp.
—— (1870) *Wiener Baedeker. Wanderungen durch Wien und Umgebungen. Mit 2 Stadtplänen, 6 Theaterplänen und 30 Holzschnitten*, 2nd edn, Wien: Faesy & Frick.
Busson, P (1913) *Wien, seine Sehenswürdigkeiten und Vergnügungen*, Illustrations F Schönpflug, J Danilowatz (Landesverband für Fremdenverkehr in Wien und Niederösterreich, Wien).
Crang, M (1996) 'Envisioning Urban Histories: Bristol as Palimpsest, Postcards, and Snapshots', *Environment and Planning A*, 28: 429–52.
Csendes, P and Mayer, W (1987) *Die Wiener Straßennamen. Katalog zur Kleinausstellung des Wiener Stadt- und Landesarchivs*, Wien: Verein für Geschichte der Stadt Wien.

Czeike, F (1983) *100 Jahre elektrische Tramway in Österreich. 80 Jahre Wiener Städtische Straßenbahn. Jubiläumsausstellung des Modellbauers Heinz Fink im Wiener Dorotheum, Wien 1, Dorotheergasse 17, 27. Juni bis 9. Juli 1983. Festschrift* (Verein für Geschichte der Stadt Wien im Zusammenwirken mit dem Wiener Stadt- und Landesarchiv, Wien).

—— (1987/8) 'Die Entwicklung des öffentlichen Verkehrs in Wien', *Handbuch der Stadt Wien*, 102: II/1–II/25.

—— (1988) 'Landpartien und Sommeraufenthalte. Die Entwicklung vom ausgehenden 17. bis zur Mitte des 19. Jahrhunderts', *Wiener Geschichtsblätter*, 43: 41–64.

—— (1989) 'Verkehrsprobleme im alten Wien. Fragen der Verkehrssicherheit im 18. und 19. Jahrhundert', *Wiener Geschichtsblätter*, 44: 1–20.

de Certeau, M (1988) *The Practice of Everyday Life*, Berkeley, CA, Los Angeles, CA, London: University of California Press.

Donald, J (1999) *Imagining the Modern City*, London: The Athlone Press.

Edensor, T (1998) *Tourists at the Taj: Performance and Meaning at a Symbolic Site*, London: Routledge.

—— (2000) 'Moving Through the City', in D Bell and A Haddour (eds) *City Visions*, Harlow: Pearson Education Limited, pp. 121–40.

—— (2001) 'Walking in the British Countryside: Reflexivity, Embodied Practices and Ways to Escape', in P Macnaghten and J Urry (eds) *Bodies of Nature*, London, Thousand Oaks, CA and New Delhi: Sage, pp. 81–106.

Fyfe, N R (ed.) (1998) *Images of the Street: Planning, Identity and Control in Public Space*, London and New York: Routledge.

Geusau, A Reichsritter von (1803) *Kurze Beschreibung der k.k. Haupt- und Residenzstadt Wien in Oesterreich, und besonders des Panorama von dieser Hauptstadt, derselben Vorstädten und umliegenden Gegenden*, Wien, Prag.

Holston, J (1989) *The Modernist City. An Anthropological Critique of Brasília*, Chicago, IL and London: The University of Chicago Press.

Jarvis, R (1997) *Romantic Writing and Pedestrian Travel*, London: Macmillan.

Kaiser-Jubiläum's-Führer durch Wien (1898) H. Bürger (ed.), Brünn: Rudolf M. Rohrer.

Koshar, R (2000) *German Travel Cultures*, Oxford and New York: Berg.

—— (2002) 'Germans at the Wheel: Cars and Leisure in Interwar Germany', in R Koshar (ed.) *Histories of Leisure*, Oxford and New York: Berg, pp. 215–30.

Kretschmer, H (1988) 'Hotelboom – gestern und heute', *Wiener Geschichtsblätter*, 43: 71–9.

Larsen, J (2004) *Performing Tourist Photography*. Unpublished Ph.D. thesis, Roskilde University, Roskilde, Denmark.

Leslie, E (2002) 'Flâneurs in Paris and Berlin', in R Koshar (ed.) *Histories of Leisure*, Oxford and New York: Berg, pp. 61–77.

Lewis, N (2001) 'The Climbing Body, Nature and the Experience of Modernity', in P Macnaghten and J Urry (eds) *Bodies of Nature*, London, Thousand Oaks, CA and New Delhi: Sage, pp. 58–80.

Meißl, G (2000) 'Hierarchische oder heterarchische Stadt? Metropolen-Diskurs und Metropolen-Produktion im Wien des Fin-de-siècle', in R Horak, W Maderthaner, S Mattl, L Musner, G Meissl and A Pfoser (eds) *Metropole Wien. Texturen der Moderne*, Wien: Wiener Universitätsverlag, pp. 284–375.

Neuester Wienerischer Wegweiser für Fremde und Inländer, vom Jahre 1797. Oder kurze Beschreibung aller Merkwürdigkeiten Wiens, der neuen Bearbeitung zweyte verbesserte, vermehrte Auflage, mit Kupfern (1797) Wien: J. Camesina.

Schivelbusch, W (2000 [1977]) *Geschichte der Eisenbahnreise*, Frankfurt am Main: Fischer.

Schmidl, A (1833) *Wien wie es ist. Ein Gemälde der Kaiserstadt und ihrer nächsten Umgebungen in Beziehung auf Topographie, Statistik und geselliges Leben, mit besonderer Berücksichtigung wissenschaftlicher Anstalten und Sammlungen nach authentischen Quellen dargestellt von A. Schmidl. Mit einem Plane der Stadt und Vorstädte.* Wien: Carl Gerold.

—— (1840) *Wien wie es ist. Die Kaiserstadt und ihre nächsten Umgebungen nach authentischen Quellen, mit besonderer Berücksichtigung wissenschaftlicher Anstalten und Sammlungen, und einem Anhange: acht Tage in Wien, als Anleitung die vorzüglichsten Sehenswürdigkeiten im kürzesten Zeitraume zu besuchen, dargestellt von A. Schmidl. Mit einem Plane der Stadt und Vorstädte. Dritte verbesserte Auflage*, Wien: Carl Gerold.

Smith, A D (1991) *National Identity*, London: Penguin Books.

Spellman, C (1996) 'Usynlige monumenter', in M Zerlang (ed.) *Byens pladser. Urbanitet & Æstetik*, København: Borgen, pp. 147–64.

Spring, U (forthcoming) 'Den severdige byen: Vandringer gjennom Wien og det kulturelle rom på 1800-tallet', in B Skarin-Frykman *et al.* (eds) *Kulturarv og komparasjon*, Oslo.

Thrift, N (2000) '"Not a Straight Line But a Curve", or Cities are not Mirrors of Modernity', in D Bell and A Haddour (eds) *City Visions*, Harlow: Pearson Education Limited, pp. 233–63.

Towner, J (1996) *An Historical Geography of Recreation and Tourism in the Western World 1540–1940*, Chichester: John Wiley & Sons.

Urry, J (1995) *Consuming Places*, London and New York: Routledge.

—— (2002) *The Tourist Gaze*, 2nd edn, London: Sage.

Virilio, P (1984) *L'horizon négatif. Essai de dromoscopie*, Paris: Éditions Galilée.

Wallace, A D (1993) *Walking, Literature, and English Culture*, Oxford: Clarendon Press.

Weidmann, F C (1859) *Neuester illustrirter Fremdenführer in Wien. Mit einem Plane der Stadt und Vorstädte. Siebente verbesserte Auflage*, Wien: Tendler & Comp.

CHAPTER THREE

Between the Physical and the Virtual: Connected Mobilities?[1]

Peter Adey and Paul Bevan

> *You've only to consider, or better still, to see and touch mountains as formed by their folding, for them to lose their solidity, and for millennia to turn back into what they are, not something permanent but time in its purest state, pliability. There's nothing more unsettling than the continual movement of something that seems fixed.*
>
> (Deleuze 1995: 157)

The city is increasingly seen in terms of flow and flux. The mobilities of people, cyborgs (Haraway 1991, 1995, 1997; Featherstone 2000), things and information extend within and beyond the usual limits of the city. Furthermore, space has become reorganized, (re)combined and permeated by technologies of extended virtual connectivity through telecommunications and ICT development. In this chapter, it is our intention to explore the development of these mobile and emergent technospaces through two specific sites within the city of Liverpool, UK. Using the empirical case-studies of an Internet website and the city's regional airport, we attempt to illustrate the connectivity of the mobilities that intersect these spaces. The chapter examines how these mobilities have become increasingly entangled in both virtual and physical worlds. We can term these *cybermobilities*: motions that transgress their (im)material domains, transmaterial movements engaged by cyborg bodies linked to complex and wide-ranging physical and virtual networks.[2]

The first of these mobilized spaces is glasswerk.co.uk, a music-based website that produces reviews, gig guides and CDs. The company is a product of an increasing regional and global emphasis on technologization where networks of connectivity and access to ICTs have encouraged the development of Internet-centric business practices. Our second case-study, Liverpool John Lennon Airport, reflects the growing development of regional airports and the cheapening of air travel. The terminal has undergone considerable redevelopment and expansion in recent years and has become the low-cost hub for the north of England.

Through these case-studies we want to explore not only how mobility can be rethought, but also the empirical changes in mobility itself. Specifically we can ask how is Liverpool becoming reworked, or re-spatialized by new mobile technologies of the city? How are physical and virtual mobilities becoming incorporated in different ways?

To begin, we briefly outline our case-studies before examining two closely running veins in urban theory. First, we explore the understanding of cities as spaces of continual and differential movement and, second, a current body of work that comprehends the city through what Graham would term 'cyber-cities' (2004), urban space that is permeated by infrastructure, technology and virtual connectivity. In following and reconciling these two directions we develop the concept 'cybermobilities'. Using our two examples we go on to examine the mobilities of Glasswerk through the connected mobility of bits and information and their relationship with the subsequent movement of people and objects. The following section goes on to explore the increasing virtuality of the newly redeveloped Liverpool John Lennon Airport terminal, where we examine how virtual mobilities work to affect the physical movement of passengers. The chapter then concludes with the suggestion that cities such as Liverpool may be remixing or recombining through these emergent and connected mobilities.

Liverpool – 'From Seaport to E-port'?

Both case-studies, Glasswerk and Liverpool Airport are located within the city of Liverpool, UK. Liverpool, a city of nearly 450,000 people has traditionally been associated with the physical forms of mobility that are tied closely to the city's famous and historic seaport. The city originally acted as an anchor point in the triangular trading system of European goods and materials exchanged for slaves in Africa, who were then transported to the West Indies and the Americas in exchange for sugar, tobacco and cotton. In the eighteenth and the beginning of the nineteenth century Liverpool became a gateway for trans-Irish Sea movements as well as transatlantic migration into and out of the UK. The country's first railway station, moreover, was constructed to connect Liverpool to Manchester, creating a region central to the British 'transport revolution' during the age of industrial expansion (on which see J. Jain, this volume). Liverpool has also acted as a nexus and hub for 'global' cultural movements. The second half of the twentieth century saw 'The Beatles' reach world-wide fame, launching Liverpool into the public consciousness through their records, radio broadcasts, touring and public performances.

The first Liverpool Airport, built in the suburb of Speke, less than ten miles from the city centre, was a huge and significant undertaking in the development of a British interwar aviation network. The development was considered as important as London's Croydon, Berlin's Templehof and Paris's Le Bourget (Butler 1983). After some decline the airport has been taken over and redeveloped, providing a base for the no-frills airline easyJet and includes others such as Ryanair. The renamed Liverpool John Lennon Airport has become a low-cost hub for the north west of England, offering flights to London, Ireland and Europe. In every sense the airport attempts to continue the city's traditional position as a node for physical movement, while also signalling the city's international cultural status through the terminal's 'John Lennon' branding.

Yet, movement within Liverpool is not limited to the physical. The steady development of telecommunications infrastructure, cable television systems and digital broadband connections, prompted first by academic and corporate needs and later by consumer demand, has allowed an increasing level of mobilization of the virtual to take place. The interconnection of these technologies through both business parks and consumer products has provided the means for companies seeking to make use of the 'e-port' to thrive. Our second case-study takes the form of one such company, Glasswerk media – a business that specializes in web-based marketing for music event promotions and gig guides.

Cities, Mobilities and the Virtual

It is not only Liverpool, but also the very concept of cities that is often equated with mobility. Rather than view the city as a static fixed thing, cities have been imagined as living organisms where streets and roads provide the circulatory systems of the living metropolis (Sennett 1994). Freely flowing mobility is the 'social barometer' or perhaps thermometer of the well being of the city (Scobey 2002), while the clogged arteries of traffic, people and waste create unhealthy problems of congestion and pollution (Marvin and Medd 2006).

Moves towards the understanding of the socio-technical side of cities through approaches from Science and Technology Studies (STS) and Actor-Network Theory (ANT) allow us to further view the hidden networks of things, of energy and water supplies, and of the virtual movement of information and data (Graham and Marvin 1996, 2001). These networked cities become gearboxes as people, things and information move at different speeds, to different places, producing social landscapes of differentiated movement, access and quality (Cresswell 2001; Hubbard and Lilley 2004; Massey 1993). In viewing the city in terms of movement, flow, flux and complexity have become the key metaphors or paradigm to view the urban (Prigogine and Stengers 1984; Clark 2000; Smith 2003; Urry 2003).

Mitchell (1996) and others have suggested that cities are being transformed by technologies that intersect the physical, the virtual and the mobile, in libraries, museums and stations. Software, according to Thrift and French increasingly writes the city (2002). Cities are becoming wired, as business and commerce must connect to not just the globalized 'elsewhere' (Allen and

Hamnet 1995), but also to the local.[3] 'Recombinant architecture', 'cybercities' and 'virtual communities' are terms used to describe the increasing bond of the urban to teletechnologies (Graham 2004; Boyer 1996; Mitchell 1996, 2003; Rheingold 1994). Whereas scholars have mapped the transport connections and amenity networks such as water and energy pipelines, we are now also able to represent the tendrils of connection to the communication networks, including the Internet (see Dodge and Kitchin 2001).

However, although the city has begun to be seen in these terms – of both physical and virtual mobilities – all too often they remain separated or distanced. This has resulted in studies that tend to view physical mobility with little consideration of its relationship with virtual movement. And at the same time studies of the virtual have the propensity to either discard or romanticize physical barriers such as the body for such virtual mobility, sidelining physical effects into pre-access difficulties or a dislocated 'meatspace' (Gibson 1984).

Recent work however, has begun to reconcile these disparate elements of the city. The urban may now be understood as an integrated complex of the physical, the virtual and mobility. The architect and theorist Marcos Novak uses the term *transarchitectures* to describe these changes – where software, telecommunications and matter increasingly colour and pollute one another (1997). Cities in this respect have become the:

> staging posts in the perpetual flux of infrastructurally mediated flow, movement and exchange . . . They are the dominant sites of global circulation and production within a burgeoning universe of electronic signals and digital signs . . . they are overwhelmingly important in articulating the corporeal movements of people and their bodies.
>
> (Graham 2004: 153)

The urban becomes both the product and the conduit of what we want to call *cybermobilities*.

Continuing this burgeoning field, work within surveillance studies has examined the identification, virtual mediation and control of car drivers through Intelligent Transport Systems (ITS) (Bennett *et al.* 2002). David Holmes' study of Melbourne's CityLink illustrates the dependence of physical mobility upon complex telecommunication systems (Holmes 2004). Furthermore, such studies have illustrated the logistical organization of cargo and goods through distribution centres, airports and goods exchanges (Easterling 1999). These examples are, for us, emblematic of cybermobilities. They not only demonstrate the electronic control and organization of physical movement, but they also reveal the equally mobile characteristics of the technologies that do this.

For us, cybermobilities are movements that are not merely existent in independent material worlds, but they are, rather, movements that blur and flood these boundaries. They not only pass through cities but they also make them, with both elements inexorably bound up and dependent upon one another. One example of this is the case of road tolling through ITS. Within these systems, the physical mobility of the car/driver is facilitated by the virtual mobility of information. This electronic movement communicates the recognized licensed number plate that is then matched with the identity of the car driver who is billed or checked against sufficient funds (Holmes 2004). In this sense, the physical component of cybermobility is entangled with parallel virtual movements, just as the virtual component of cybermobility is equally dependent upon the material movement – of person, or car. Cybermobilities are therefore mutually dependent and necessary, similar to the manner that Dodge and Kitchin suggest a 'code/space' may result from the dyadic and 'mutual constitution of code and space' (2004: 209). In this sense, cybermobilities are the combination of virtual and physical movements, which are symbiotically produced and mediated through one another.

We can now turn to our case-studies to draw out other examples of these cybermobilities. We discuss the physical mobilities of people and things that intersect with the virtual movements of a website, and from the other side of the spectrum, the virtual movements that mediate the physical within an airport.

Glasswerk

In many senses, glasswerk.co.uk functions as a static location that allows the elements of the company to retain a 'brand unity' and provide a point of access for the business. The apparent fixity of the site is, in fact, the coming together of various networks that are themselves in constant motion. The production of content is spread between music venues, sites of writing and the editorial office in Wavertree. Similarly, this content – once it has been translated into the code from which the site is rendered – is stored not in Liverpool but in the server arrays of a Cambridge, UK, hosting company located in Houston, Texas.

Glasswerk exists in a fluid suspension, perpetually between points of presence, which intermittently coalesce into the performance of the website. The pathways of data between users, infrastructures and servers reflect the mobility of the site, for without the mobility of this code there can be no 'whole', or representation, of the Glasswerk website. This necessary mobility, a constant flux essential to the appearance of obduracy, could be termed non-representational (Thrift 2000) as its continual 'performance' is inherently bound up in the mobility of all its elements. To an extent, the city has become what John Rajchman (2000) describes as 'invisible and unlocalizable', as it is distributed across the elements of its networks. But perhaps he goes too far when he writes:

> electronic space through which we move and make moves 'exposes' the city to something that can no longer be read as a structuring or framing network or seen through the materials and networks that realise it . . . it no longer requires the sort of physical displacements that provided the sense of mobility and congestion captured in the progressive futurist imagination.
>
> (Rajchman 2000: 29)

In contrast to Rajchman's suggestion, the site's mobility is also wrapped by a cyborgic movement of information between the material and the virtual, with on and offline becoming entangled in transmaterial effects. In this section we will explore two particular examples of how Glasswerk intermediates the physical and the virtual, providing the means and the motive for these cybermobilities.

Virtualized Performances

Music events form the basis for glasswerk.co.uk's business. Without the material performances that take place either under the guise of the company or independently, the site would have no content. In fact, Glasswerk's principal function is to act as a conduit between the material and the immaterial. In this section we look at the ways in which this connectivity produces, or enhances, the mobility of both musical performance and music devotees.

While Glasswerk has become an important resource and discussion forum for unsigned bands, the original intention was to provide a way of promoting the music of those bands that performed under the 'Glasswerk presents' label. Actual sales of CDs of these bands are minimal, and a wider role is played by digital formats, which allow quick and easy access to the music. Mat Ong, the CEO of the Glasswerk group, illustrated this by saying:

> The CD sales, as I say, come from abroad so we're just basically . . . a distributor of music. The main thing is the MP3s which means people can listen to the music. I don't really like selling CDs that much 'cos it's a big hassle 'cos you have to make them, and post them and it's all this thing – you don't make any money off it. We do that because . . . just as a bonus for bands because it gives them the option to sell music on the Internet but it's not a major factor.
> (Interview with Mat Ong, 28/08/2003)

The physical pathways along which CDs must travel slow down the mobility of this music, while at the same time providing a tangible, and desirable, service for the bands involved. Digital media, in this case in the MP3[4] format, can be stored once on the site and downloaded countless times without any degradation of quality. The encoding of live, and physical, gigs and sessions to digital formats such as MP3 provides the means for bands' music to move from the physical mobility of sound waves, gig venues and live music to the easily exchanged commodification of code.

The prolific use of digitized video and audio on the Glasswerk site also offers increasing opportunity for linkages between artists. Whereas live gigs hosted by Glasswerk seek to encapsulate particular musical tastes through selected line-ups, with bands jostling for the coveted 'final spot', the website allows easy movement between bands – and their digitized recordings – according to individual taste, momentary whim or serendipity. These connections provide the potential for an intermeshed musical journey and as such increases the audience of particular bands (whether online or followed up in physical form). This symbiosis was a direct consequence of the way in which the site was conceived:

> PB: *I really got the impression [from the business plan] about how*
> *you and the artist could promote each other's brands ... by*
> *offering the opportunity for them to listen to their music they're*
> *coming back to your site and listening to other artists as well?*
>
> MO: *That was the idea when we first started, we didn't have a big*
> *budget when we started up – well we started with no contacts,*
> *we started with completely nothing – we knew nothing about*
> *anything and we didn't know anybody ... I guess the idea was*
> *the bands on the Glasswerk thing they would look at other bands*
> *and they would send their mates to go and look at Glasswerk*
> *as well so it built up a following quite quickly – just by doing*
> *that kind of word of mouth kind of thing.*
>
> (Interview with Mat Ong, 28/08/2003)

Mobilized Events

The way in which bands are linked through the Glasswerk site – a portal if you will between the physical performance of music and the collation of digital resources and discussion forums – is interrelated with the second form of mobility we will discuss in this section, the physical mobility of gig-goers.

It is important to bear in mind that the Glasswerk site is viewed by the company as essentially a very complex marketing tool for live performances. Very little money is made through the website; however, a sizable investment in infrastructure – both in the form of hardware and Internet connectivity – and in time is needed to keep the website running. Rather than being simply a drain on resources, however, glasswerk.co.uk acts as a central pivot between material performances and the immaterial garnering of information by their potential audiences.

The branding of gigs under the banner 'glasswerk.co.uk' connects the live performance with the website and thus increases movement online – both in terms of prior and post-event research – so undercutting the hypothesis of 'substitution' (Plaut 1997). 'Hits', or page views, will increase following performances and many of those users will return in the future to see what else is promoted through the site. In this way just as online mobility is influenced by attendance at offline, material, performances – or even the use of the glasswerk.co.uk moniker on posters and event cards – attendance at future gigs becomes influenced by the information available on the website. Thus, the physical mobility of music devotees becomes increasingly bound up with the virtual mobility they experience online.

Glasswerk has, by a function of its deep linkages between online marketing and offline performances, blurred the boundaries between the material and the immaterial, creating fluid, cyborgic, cybermobilities. The company itself straddles this complex transmateriality in both its content and diverse located-ness (following Graham 1998). This interrelatedness is a central element of the company, transgressing norms of practice to the extent that when the CEO of Glasswerk was asked how he saw the site's relation between the local and the online he struggled to express the links between the company's role in both the website and the performances:

> MO: *That's the whole point – it's like an online company . . . but it's not an online company it's a real-world company. But it has shifted from the original concept of being a company . . . doing some real-world stuff to be a real-world company with a high-level of online presence.*
>
> (Interview with Mat Ong, 28/08/2003)

Glasswerk's money-making events are embedded in the background of Liverpool's musical history. By virtue of this, the integral relationships between the virtualized and performed elements of Glasswerk result in Liverpool itself being mobilized through the virtual network of the Glasswerk website. Glasswerk's example is also a theme that can be mirrored throughout the city. Liverpool City Council has developed a £300 million partnership with British Telecom to update and outsource its call centres and information points and to integrate existing infrastructures. This movement from 'seaport to e-port' aimed to virtualize and speed communications between the authority and the public (Liverpool City Council 2001). Online developments, electronic city centre access points, and the possibilities of other mediums such as mobile telephony and Digital TV are proposed to allow residents other doorways to council services and local information. According to the council's rhetoric, the city is undergoing an 'e-volution'.

Initiatives such as the 'Interactive City' conference that took place in June 2004, bringing together industry leaders in information communications (O'Callaghan 2004), should also be read alongside the increasing success of software gaming industries within the city. Such industries are proposed to be developing due to the traditional geographical clustering of cultural and creative talent in the city along with the improvements in communications hardware (O'Callaghan 2003). The growth of a large number of call centres is also another example of this trend.

Liverpool John Lennon Airport

Rather than examining airports in electronic and virtual terms, what can be labelled traditional approaches have used airports to symbolize the networked material fluidity of globalization and post-modernity (Ibelings 1998). Other sociological approaches are exemplified by Marc Augé's *Non-places* (1995; see also Castells 1996). Augé's lexicon identifies these spaces as non-places – spaces empty of any meaning and social significance. For Augé, the mobility of people defines the transitory nature of these spaces and the fleeting presence of attachment (see Crang 2002). Airports are also seen as complexes of material movement that mobilize and channel thousands of workers, and the flows of goods, cargo and baggage. The development pioneer H. McKinley Conway writes that where urban growth was first encouraged by the railroad and the highway, the airport persuades another centre of development – the airport or 'fly-in' city, where 'new patterns are unfolding as attempts are made to make new activity centres near runways' (1978: 5). It is impossible for the airport 'to escape from the city' (1978: 239).

Yet the physical space of the 'airport city' is beginning to be understood by the vitally important electronic and telecommunication infrastructures that airports intersect. This is in addition to the software code that knits the airport's complexities together (see also Adey 2004a, b; Dodge and Kitchin 2004; Gottdiener 2001; Pascoe 2001). For example, in their study of Changi Airport, Singapore, Rowley and Slack (2002) demonstrate how mobile communication technologies and Internet access have inundated the space. The airport has become the 'gateway to Singapore's knowledge society' (Slack *et al.* 2004). Virtual mobilities take an even more important role for Dodge and Kitchin (2004) who argue that the airport is a 'code/space'. They mark airports as distinctive from other 'coded spaces', for within the system of air-travel and airports, software is absolutely essential. Take, for instance, how passengers' luggage and cargo are sorted and logically organized by software code (Easterling 1999). Aircraft are also dependent upon their communications between airports and air traffic control centres (Wells 1996).

In terms of basic passenger numbers, the airport is dwarfed by the neighbouring international airport at Manchester; yet this belies the incredible growth of Liverpool Airport in recent years, which reflects the economic outlook of the city as a whole. With considerable investment from its parent company, Peel Holdings Ltd, the airport was able to develop and refurbish the 1980s' terminal building and attract names from the low-cost airline market such as easyJet and Ryanair. The airport has also been accredited with the title of the fastest growing airport in the UK, growing from around 1 million passengers in 1999 to almost 3.5 million in 2004. However, given the considerable differences between any airport, both in size and complexity, it is not surprising that there are resulting implications upon the communications networks and technologies that an airport may employ. It is then important not to paint all airports with the same brush. The complete virtualization of Liverpool John Lennon Airport is not nearly as essential as somewhere such as Manchester, Gatwick or Heathrow as perhaps a 'code/space' may imply (see Dodge and Kitchin 2004). That said, many of the systems that are in place are crucial to the safe and smooth running of the airport.

Furthermore, in filling the terminal with a raft of low-cost airlines, Liverpool provides an alternative to Manchester's traditional flag-carrying and charter traffic. Direct competition with Manchester may also not be too far away. Plans are already under way to bring forward the airport's development schedule now that Ryanair and Flybe will begin to use Liverpool as a northern base. The development of the low-cost airlines is also significant in terms of the virtualization of air travel. Because many of the low-cost airlines work upon a business model of Internet and telephone bookings, the passengers' first entry point to the airport is often through a computer screen or a phone – before their physical journey has even begun.

Information

One of the most common forms of cybermobilities at the airport is the inter-section between the virtual mobility of information and the material movement of passengers. The Liverpool Airport terminal itself is interwoven with tele-communication infrastructural networks. To avoid information overload the air-port is attempting to develop into a silent airport, therefore the majority of information that the passenger receives comes from the signage system, informa-tion desks and the flight information displays (FIDS).[5] The FIDS highlight the most obvious form of virtual movement within the terminal. Software such as the Airport Management Operational Support System (AMOSS) manages the movement of information through the various communication infrastructures that overlay and interweave the terminal building, from computer terminals, to radar and mobile telephony such as SMS messaging. The system interprets and permits information to be quickly passed and shared between the airport community, and displayed on the television screens.

The textual information displayed on these screens provides part of the informational architecture of the airport, which, according to Fuller has become: 'a highly textually mediated space where the signs not only enact semioticised territories but also directly intervene in the material machinic processes of travelling' (2002: 133). As opposed to signage, this form of communication is obviously far more adaptable than standard signage systems and contributes to the increasing reflexivity of these environments. The mobility of this informa-tion to the passenger is vital if the strict turnaround times necessary for low-cost travel are to be achieved. The screens direct passengers to their gate, they tell them where to wait, and when to go.

This is not to say that this information is limited to the airport terminal. At Liverpool Airport the vectors of communication extend far beyond the terminal doors. The newly introduced AMOSS system means that the informa-tion displayed on the FIDS is a combination of not only scheduled data, but data it is receiving from incoming and outgoing flights as well as ground move-ments. When a flight leaves an airport it sends out a signal that is received by the destination airport, containing information concerning whether the flight has left on time or whether it is late. Combined with other real-time data the FIDS then display suitable times for passengers to pass through security pinch-point, and make their way to the departure lounge. These informational move-ments inform and direct passengers' real-time mobility through the terminal space (see J. Jain, this volume, on real-time train scheduling). The airport has become a world: 'where the informational and the material increasingly stream through each other' (Fuller 2002: 142).

Simulation

Another growing aspect of this development is the use of modelling simulations in airport architecture (Adey 2004b). For the architectural historian, Charles Jencks, software simulations and computer-aided design (CAD) have evolved a 'new paradigm' in architectural thought (Jencks 2002). Traditional building designs have been 'warped' by these changes in practice as the paper and ink design has left the material drawing board and moved instead from 'disk to product' (Vidler 2001). According to Marcos Novak: 'Computers have created a direct link between what can be conceived and what can be built, augmenting both' (Novak 2001). At many airports, simulations have therefore become the primary tool to enable airport planners and managers to predict changes made to the space. The potential impacts upon passenger movement and the efficient running of the space can be studied with regard to Liverpool John Lennon Airport.

Simulated virtual movement through the airport was an integral part to the design of the newly completed terminal. The architects Leach, Rhodes Walker built several 3-D walk-throughs and a fly-through of the terminal (see Figure 3.1). Through 3-D simulation the airport planner and airport management were able to experience the airport from the embodied visual perspective of a passenger. As a tool of communication this form of virtual mobility allowed the financiers of the project to effectively see the airport before it was built. Architects, planners and financiers could therefore avoid costly mistakes and changes to the building once it reached its physical form.

Simulation also allowed airport planners to predict how passengers walking through the building would perceive the airport, again avoiding costly changes and mistakes. This virtual movement of virtual passengers also became instrumental to the management of real passengers. Passenger movement must be carefully controlled to allow efficient and smooth flows through the terminal space. Sufficient numbers of passengers must also circulate around retail areas of the terminal space because low-cost airlines have meant that airports such as Liverpool must regain income from retail rent and car parking. In order to ensure the sufficient usage of these outlets, as well as efficient movement from landside to airside, the architecture of the airport is planned to carefully shape passenger behaviour. The virtual mobility of passengers within the 3-D simulations was then used to demonstrate the most effective behaviour of real passengers. The physical space of the airport is then constructed in line with the 3-D virtual model so that the building may materially shape these movements.

Although the simulations are clearly used prior to passengers moving through the space, they nonetheless play an important role. If we take Hetherington's proposal: 'The absent can have just as much of an effect upon relations as recognizable forms of presence can have' (Hetherington 2004: 159),

Figure 3.1 *Architect's Simulation of Passenger Movement*
Source: Reproduced with permission of Leach Rhodes Walker Architects and Liverpool John Lennon Airport.

in the case of the discussed simulations we can see that the 3-D model has an absent-presence in the terminal. The guiding principles and assumptions of the model do not directly affect the passenger, but instead become inscribed and deferred into the material structure of the building. The logic of the simulation is *transmitted* into the physical space (Novak 1997).

Mobility, Moorings and Cybermobilities

In this chapter, we have discussed new forms of movement that work between both physical and virtual worlds. Rather than viewing mobility as something that is isolated to either material or virtual domains, our two case-studies focused upon the space of a website and a regional airport have attempted to draw out the connecting interaction between movements of both the material and the immaterial. We have argued that these mobilities may be deemed cyber-mobilities.

Where Graham observes that '[e]very physical movement has its compu-terised trace' (Graham 2004: 113), we would argue that this trace is inevitably a virtual *movement* that has affected, or has entailed, a physical movement. And, on the other hand, it is becoming inevitable that physical transport through the city will also affect or be affected by a virtual movement. As the virtual becomes increasingly interwoven in space, it seems ever more likely that this trend will continue. But if we are correct and these cybermobilities are progressively taking shape, how does this concern conceptions of stability and immobility? If we take Urry's mobility/moorings dialectic, he explains:

> This relationality between mobilities and immobilities is a typical complexity characteristic. There is no linear increase in fluidity without extensive systems of immobilities. Thus the so-far most powerful mobile machine, the aeroplane, requires the largest and most extensive immobility, of the airport city employing tens of thousands of workers.
>
> (Urry 2003: 125)

According to Urry (2003), the spaces of pause and rest facilitate the mobility of people, things and information. This relationality of mobility, or what Massey (1993) terms a 'politics of mobility', requires that someone or something moves at the expense of another's immobility (see also Cresswell 2001).

And yet, we would argue that the moorings that allow and constrain mobilities, are also becoming mobile and may operate across (im)material realms and distinctions. In the case of the discussed cybermobilities, both virtual and material movements provide the mooring for each other's passage. Therefore, as we discussed through the examples from Glasswerk, the mobility of bits and information that make up the website allows the site to be used and to function for people to attend gigs. Taking our other example of an airport, airports are probably the most powerful technology of physical mobility in the world today. Yet, it is the mobility of information that provides the mooring from which passengers and aircraft may push off.

Perhaps it is the quickness or slowness, or even invisibility of these movements that leads them to appear static and unknowable. Whether they enable or constrain other movements, a pause or rest, is really only in relation to something else's quicker speed or different direction of movement. Mobilities are better envisioned as a series of mobility differentials of speed, direction and meaning (Hubbard and Lilley 2004). Consequently, less static metaphors can be more useful than traditional notions of stability and immobility and may lead to more fruitful explorations of differential and hybridized cybermobilities. Manuel De Landa (1997) provides just such a solution in his history of the mobility of energy, matter and language:

> if we consider that the oceanic crust on which the continents are embedded is constantly being created and destroyed (by solidification and remelting) and that even continental crust is under constant erosion so that its materials are recycled into the ocean, the rocks and mountains

that define the most stable and endurable traits of our reality would merely represent a local *slowing down* of this flowing reality. It is almost as if every part of the mineral world could be defined simply by specifying its chemical composition and its *speed of flow*; very slow for rocks, faster for lava. Similarly, our individual bodies and minds are mere coagulations or declarations in the flows of biomass, genes, memes, and norms.

(De Landa 1997: 258)

De Landa is also echoed in the words of Serres and Latour who write: 'one must concede that everything is not solid and fixed and that the hardest solids are only fluids that are slightly more viscous than others' (Serres with Latour, cited in Smith 2003: 569).

Perhaps we can understand the electronic and physical restructuring of Liverpool in terms of these mixtures, a city of 'transitory hardenings and fluids' (Smith 2003: 569), where the viscosity of one mixture constrains or eases the flow of another. Cybermobilities – connected movement that inhabits and inscribes both virtual and physical space simultaneously – exemplifies this fluidity and *milieu*. What we have shown through this chapter is perhaps a remixing of the mobilities of Liverpool as the informational and physical transport infrastructures have been reworked from relatively separate realms. Traditionally the city has operated as a nexus of physical movement par excellence, supporting the flow of goods, people, slaves, trains and cargo. But it has become a city infused with both physical and virtual mobilities, a mesh of portals that link on and offline in blurred and hybridized performances. Through this conceptualization, we have shed light on both the interaction and interdependence of post-modern mobilities – reworking traditional concepts in order to (re)envision such movements as continual, connected and prevalent – as cybermobilities.

Notes

1 An earlier version of this chapter was presented at the Alternative Mobility Futures Conference held in Lancaster, January 2004. We would like to thank the participants of the conference and the editors of this book for their valuable questions, comments and suggestions.
2 We look upon transmateriality as the movement, or constant flow, between the physical or material and the virtual, or immaterial. We would argue that this is not necessarily a function of our increasing relationship with technology – although that itself is inherently transmaterial – but that transmateriality is evident in the constant process of mind and body, the virtualization of images seen through the eyes or the materialization of thoughts on to paper in writing (Bevan 2002).
3 Although it has been noticed that the most powerful and speedier links are to the elsewhere, bypassing the local and the nearby (Graham and Marvin 2001).
4 Motion Picture Experts Group audio layer 3 (or MP3) is a method of encoding audio whereby some elements of the signal are discarded in order to reduce the file size. MP3 is also used as a generic descriptor for many different kinds of digital audio as well as

for illegal file trading activities. The format has so far resisted attempts to incorporate Digital Rights Management (DRM) and as such MP3 files are not commonly used in online music stores such as Apple's 'iTunes' and Coca-Cola's 'CokeMusic'.
5   The information discussed was developed from interviews with Liverpool John Lennon Airport, Leach Rhodes Walker and FS Walker Hughes over a period between 2003 and 2004. Thanks also to the Friends of Liverpool Airport for their discussions and support.

Works Cited

Adey, P (2004a) 'Secured and Sorted Mobilities: Examples from the Airport', *Surveillance and Society*, 4 (1): 500–519. http://www.surveillance-and-society.org.
—— (2004b) 'Surveillance at the Airport: Surveilling Mobility/Mobilising Surveillance', *Environment and Planning A*, 36 (8): 1365–1380.
Allen, J and Hamnett, C (1995) *A Shrinking World: Global Unevenness and Inequality*, Oxford: Oxford University Press.
Augé, M (1995) *Non-Places: Introduction to an Anthropology of Supermodernity*, London: Verso.
Bennett, C, Raab, C and Regan, P (2002) 'People and Place: Patterns of Individual Identification Systems within Intelligent Transportation Systems', in D Lyon (ed.) *Surveillance as Social Sorting: Privacy Risk and Digital Discrimination*, London: Routledge.
Bevan, P (2002) 'Traversing the Materiality Divide: Gamers as Asynchronous Cyborgs'. Paper presented at The Salford Seminar on Immateriality, 8/04/2002, Salford, Manchester.
Boyer, C (1996) *Cybercities: Visual Perception in the Age of Electronic Communication*, Princeton, NJ: Princeton Architectural Press.
Butler, P (1983) *An Illustrated History of Liverpool Airport*, Liverpool: Merseyside Aviation Society.
Castells, M (1996) *The Rise of the Network Society – Volume 1: The Information Age: Economy, Society and Culture*, Oxford: Oxford University Press.
—— (2001) *The Internet Galaxy: Reflections on the Internet, Business, and Society*, Oxford: Oxford University Press.
Clark, N (2000) '"Botanizing on the Asphalt"? The Complex Life of Cosmopolitan Bodies', *Body and Society*, 6 (3–4): 12–33.
Crang, M (2002) 'Between Places: Producing Hubs, Flows and Networks', *Environment and Planning A*, 34: 569–574.
Cresswell, T (2001) 'The Production of Mobilities', *New Formations*, 43: 11–26.
De Landa, M (1997) *A Thousand Years of Non-Linear History*, Cambridge, MA: Zone Books.
Deleuze, G (1995) *Negotiations*, New York: Columbia University Press.
Dodge, M and Kitchin, R (2001) *Mapping Cyberspace*, London and New York: Routledge.
—— (2004) 'Flying through Code/Space: The Real Virtuality of Air Travel', *Environment and Planning A*, 36 (2): 195–211.
Easterling, K (1999) *Organization Space*, Cambridge, MA: MIT Press.
Featherstone, M (2000) 'Post-Bodies, Aging and Virtual Reality', in D Bell and B Kennedy (eds) *The Cybercultures Reader*, London: Routledge.
Fuller, G (2002) 'The Arrow – Directional Semiotics: Wayfinding in Transit', *Social Semiotics*, 12 (3): 131–144.
Gibson, W (1984) *Neuromancer*, London: Gollanz.
Gottdiener, M (2001) *Life in the Air: Surviving the New Culture of Air-Travel*, London: Rowman & Littlefield.
Graham, S (1998) 'The End of Geography or the Explosion of Place? Conceptualizing Space, Place and Information Technology', *Progress in Human Geography*, 22 (2): 165–185.
—— (ed.) (2004) *The Cybercities Reader*, London: Routledge.

—— and Marvin, S (1996) *Telecommunications and the City: Electronic Spaces, Urban Places*, London: Routledge.

—— and —— (2001) *Splintering Urbanism: Networked Infrastructures, Technological Mobilities and the Urban Condition*, London: Routledge.

Haraway, D J (1991) *Simians, Cyborgs, and Women*, London: Free Books Association.

—— (1995) 'Cyborgs and Symbionts: Living Together in the New World Order', in C H Grey, H J Figueroa-Sarriera and S Mentor (eds) *The Cyborg Handbook*, London: Routledge.

—— (1997) *Modest_Witness@Second_Millennium.FemaleMan©_Meets_OncoMouse™: Feminism and Technoscience*, London: Routledge.

Hetherington, K (2004) 'Secondhandness, Consumption, Disposal and Absent Presence', *Environment and Planning D: Society and Space*, 22: 157–173.

Holmes, D (2004) 'Cybercommuting on an Information Superhighway: The Case of Melbourne's CityLink', in S Graham (ed.) *The Cybercities Reader*, London: Routledge, pp. 173–178.

Hubbard, P and Lilley, K (2004) 'Pacemaking the Modern City: The Urban Politics of Speed and Slowness', *Environment and Planning D: Society and Space*, 22 (2): 273–294.

Ibelings, H (1998) *Supermodernism: Architecture in the Age of Globalization*, Rotterdam: NAI.

Jencks, C (2002) *The New Paradigm in Architecture: The Language of Postmodernism*, New Haven, CT: Yale University Press.

Liverpool City Council (2001) *Driving up Standards*, Liverpool City Council Annual Report 2000/2001, http://www.liverpool.gov.uk/pdfs/Driving.pdf (Accessed 17/11/04).

Mckinley Conway, H (1978) *The Airport City: Development Concepts for the 21st Century*, Atlanta, GA: Conway Publications.

Marvin, S and Medd, W (2006) 'Metabolisms of Obesity: Flows of Fat across Bodies, Cities and Sewers', *Environment and Planning A*, Special issue on Mobilities and Materialities (in press).

Massey, D (1993) 'Power Geometry and a Progressive Sense of Place', in T Bird, B Curtis, T Putnam and G Robertson (eds) *Mapping the Futures: Local Culture, Global Change*, London: Routledge.

Mitchell, W (1996) *City of Bits: Space, Place and Infobahn*, Boston, MA: MIT Press.

—— (2003) *ME++: The Cyborg Self and the Networked City*, Boston, MA: MIT Press.

Novak, M (1997) 'Transmitting Architecture: The Transphysical City', *Ctheory*, http://www.ctheory.net/articles.aspx?id=76 (Accessed 08/05/04).

—— (2001) 'Liquid~, Trans~, Invisible~ : The Ascent and Speciation of the Digital in Architecture. A Story', http://www.mat.ucsb.edu/~marcos/Centrifuge_Site/MainFrame Set.html (Accessed 08/05/04).

O'Callaghan, B (2003) 'Pixel Power', http://www.bbc.co.uk/liverpool/culture/2003/11/gaming/index.shtml (Accessed 17/11/04).

—— (2004) 'From Seaport to E-Port', http://www.bbc.co.uk/liverpool/capital_culture/2004/06/ic04/index.shtml (Accessed 17/11/04).

Pascoe, D (2001) *Airspaces*, London: Reaktion.

Plaut, P O (1997) 'Transportation-Communications Relationships in Industry', *Transportation Research A*, 31 (6): 419–429.

Prigogine, I and Stengers, I (1984) *Order out of Chaos*, New York: Bantam.

Rajchman, J (2000) *Constructions*, Boston, MA and London: MIT Press.

Rheingold, H (1994) *The Virtual Community: Homesteading on the Electronic Frontier*, New York: Harper Perennial.

Rowley, J and Slack, F (2002) 'Online Kiosks: The Alternative to Mobile Technologies for Mobile Users', *Internet Research*, 12 (3): 248–257.

Scobey, D M (2002) *Empire City: The Making and Meaning of the New York City Landscape*, Philadelphia, PA: Temple University Press.

Sennett, R (1994) *Flesh and Stone: The Body and City in Western Civilisation*, London: Faber & Faber.

Slack, F, Rowley, P and Rowley, J (2004) 'Keeping Customers on the Move Connected: A Case-Study of Singapore Changi Airport', *BCS Review 2004*, http://www.bcs.org/BCS/review04/articles/mobilecomputing/keepingcustomersconnected.htm.

Smith, R G (2003) 'World City Topologies', *Progress in Human Geography*, 27: 561–582.

Thrift, N (2000) 'Afterwords', *Environment and Planning D: Society and Space*, 18: 213–255.

—— and French, S (2002) 'The Automatic Production of Space', *Transactions of the Institute of British Geographers*, 27: 309–335.

Urry, J (2003) *Global Complexity*, Cambridge: Polity Press.

Vidler, A (2001) *Warped Space: Art, Architecture and Anxiety in Modern Culture*, Cambridge, MA: MIT Press.

Wells, T (1996) *Airport Planning and Management* (3rd edn) New York: McGrawHill.

CHAPTER FOUR

Urban Violence: Luxury in Made Space

Sarah S. Jain[1]

In American cities, cars and the urban form simply cannot be conceived in isolation of one another. Since the turn of the last century, the "public" spaces of the street have been accessible virtually only by car, and the hegemonic commodity fantasy consistently positioned through advertising equates the quality and desirability of cars with space taken – and fast. Both materially and representationally, access to public space requires private conveyance. In that sense car culture offers a synecdoche of capitalism more generally, in which commodities are posed as ways to socially and physically enhance the agency of individuated consumers.

One could cite numerous advertisements to support auto industry promises of space-taking. Indeed, in recent advertising a new level of individuated aggression seems to be appearing. This violence is not bound by the advertising images, but has overflowed to the street space through aggressive vehicle designs and driving practices, which inform in ever more homogeneous ways how American urban spaces are occupied by vehicles to exclude pedestrians, cyclists, and other potential road users. The violence of urban space is made manifest in the hundreds of tiny and larger intentional and unintentional aggressive moves, which fester uneasily below the radar of policy makers and cultural analysts.

It is this convergence among violence, automobiles, and the American urban form that I seek to better understand here. My primary vehicle for the analysis of this problematic will be the six-minute film, *Star*, directed by Guy Ritchie, starring his wife, and available only – but freely – on the BMW website along with a series of other advertising films with high-profile directors. This film, according to the write-up on the website, pits Madonna against the driver of the BMW (Clive Owen) in a battle of wills. The storyline is roughly this: Madonna plays the role of an extremely unpleasant if beautiful "superstar." When she insists on being taken to the venue in the BMW, rather than her limo parked nearby, Owen takes the opportunity to beat her with the car by driving fast and literally tossing her (or rather her stuntwoman) around the backseat.

The film offers many layers of artifice: the car, with its utterly dominated domestic interior; the fantastic journey across the city; the other vehicles readily giving way to the expertly driven BMW; the celebrities in the film played by real-life celebrities; and the perfected transgressive luxury of the BMW brand. The fantasy is one of escape, but the escape is not only about the car and its associated identities, it is about the places constructed by and through the car. Thus, I will use a reading of this film to argue that the car as commodity stands in a recursive relationship with the city: American urban spaces are not only built for the car, but are virtually only habitable through the car. Luxury is deeply imbued in this mechanism on the one hand as the ideal and motor of consumption (social hierarchies – something to strive for), and on the other as the technological limit of how space can be inhabited. Thus, I argue that how the car works as a commodity, and its particularity through its materiality as a *mobile* commodity, is necessary to understanding the US urban form.

I make this argument through two more detailed readings. First, *Star* offers an interpretive venue through which to better analyze the spatial and epistemic relations among film, driving, and urban space, on the one hand, and among speed, violence, and representation, on the other. I argue that while car and urban form materially produce each other in very obvious ways, the conditions of possibility for the particular way in which this has happened in the US are underlain by logics of moving images, luxury, and celebrity that mutually and compatibly produce each other. Neferti Tadiar has demonstrated in discussing the urban form of Manila, that flyovers

> constitute a particular social order. . . . Not only do they . . . represent, they *are* a system of representation: a medium. . . . [T]he metropolitan form . . . is a mode of regulation and control but also a medium of desire which helps to produce the effect *of* subjectivity.
>
> (Tadiar 1993: 155)

Here, I bring these orders of representation together further, in thinking about the material compatibility of the filmic medium and the car journey through an archive of visual culture of twentieth-century car culture films. Together, these have created certain possibilities for journeying through the city. I will suggest here that these possibilities rest on and render natural, class, gender, and mobility violence, and that *Star*, in the fullness of its self-irony, can help to render these visible. The production of the effect of subjectivity is continually and recursively consolidated through experiences of the city, roads, and visual culture. As BMW helps make clear, corporate interests have a hand in individuating these experiences and offering modes of transgression.

Second, I want to examine the ways in which the car both distributes and naturalizes these unequal "takings" of space.[2] To offer one example, car safety works within a rhetoric that at once provides protection from the unnamed public menace of the other dangerous driver at the same time as it offers the means to over-ride the public inhabitation of these spaces. Woven within these takings are politics – too often left on the level of statistics – of physical violence, life chances, and risk.

Commodities such as cars have institutional and semiotic support that render certain aspects of them visible and the relations they spur as collateral to their materiality or the so-called attitudes that are supposed as intrinsic. "Every American wants a big car" was a policy mantra of the 1950s and 1990s, inter-rupted only by a conscious effort on the part of former President Carter to make car size into a national security, rather than identity, issue. As I have argued in detail elsewhere, sometimes the acceptable damages of a commodity are class, race, age, or gender based (Jain 2005, 2006), and at other times they spur new kinds of social relations, such as they have with the introduction of sport utility vehicles (Jain 2004). But the policy mantras and statistics leave too much unsaid about humans and technologies.

The Film[3]

The film runs six minutes and 56 seconds, and is rated PG. We begin with a close-up of Owen sitting at the wheel and shown from the hood (or bonnet), a view only available through the camera's eye, and one that makes available a full-framing of the portrait in the car's black luxury interior. A lush black leather interior surrounds Owen and a seatbelt indicates his connection with – literally, attachment to – the car. Owen has the first and last word, and he begins the narrative in a not-unfamiliar deconstruction of "the star" (Madonna is not named in either the notes or the film, but she is clearly playing a version of "herself"). "The first thing you notice physically about this lady . . .," he notes, is her eyes. As the camera cuts to her sunglasses, he notes that "it is rare to actually see those eyes, 'cause they're usually covered up – but when you do" (a pause and an appreciative nod), "it is worth it." Then a cut to her hands as he tells of her "strong, powerful yet feminine" hands, and then of her voice. The voice is the ambivalent point: with it she has reached "giddy heights" and is "unrivalled in her world" – it is also the medium of her unpleasantness. As he recites this paean, the camera follows her descending in an elevator to a basement garage. She is flanked by two large bodyguards, who look to be of South-East Asian descent, donned in matching black suits and white shirts. Arriving at the parkade, she walks towards a camera set below with men dressed in black, cars, and white concrete pillars: a familiar scene from any shopping mall, university, air-port parkade, or, equally, a scary scene in many American films. She wears very dark glasses, a thick gold neck chain and ear-rings, a black tee-shirt, black gloves, short leather jacket with "superstar" bejeweled on the back, and bright red trousers. Surely she is packing something in the crotch fold of the pants: an area that will become a theme of the film. As the viewer gets a first peek, Owen completes his dismemberment of her: "She is a complete cunt."

Clipping the last letter of "cunt," the film cuts back to her. This is where we are introduced to her manager, Glen, who is described by Owen as having no "backbone," and who we see being mouthed off by the Star: "You are such an idiot," she says, her notoriously bad-acting on high, "This is not what I pay you for." "We'll work it out, sweetheart," he says. Madonna opens the back door to her limo and fairly screams, "Coffee, I want my coffee." The only other woman of the film, a brunette dressed in tight black, hands her an un-branded cup. The Star responds icily, "It better not be cold." Then, claiming to be so sick of black, she walks over to the silver BMW with Owen sitting inside. They converse briefly, he, cool, the only one not swarming around her eager to please, mentions that he is reserved for someone else. This is where the fantasy must start, since for a star of Madonna's stature everything about the trip – including the name of the venue would have been invisible. A bodyguard leans in the window to tell them the performance is at the Palace, and the Star warns him not to get into the car but to "take the bus." He jumps into a large black (bus-like?) SUV to tail Owen. She is eager to lose the trailing car, and as they leave the garage, fans pile on to the SUV. As she nags Owen to drive faster, Owen says, in his understated Gosford Park butler tone, "Well, marm, I wouldn't like to put you in any danger." She says, "Don't ma'am me" – and just then the phone rings. It is Glen: "Is she doing okay?" "Yeah, she is fine, I'll take care of her." "No rush, show her the sights, give her everything I pay you for, break-fast, lunch and dinner." Straight-faced Owen: "She'll be there on time," and glances in the rear-view. As he puts down the phone from this paternalistic wink and nudge conversation, Madonna harangues: "If you keep your eyes on the road instead of me, we might be getting somewhere."

"Let me see what I can do," he responds. And after a critical pause, "Sir."

The camera flicks back to her applying lipstick in the backseat, and the adventure begins as they peel out of the intersection. Rock music begins, Blur sings "Song 2," with lyrics that include many repetitions of "woohoo." The soundtrack also features screeching brakes, the screaming woman, and the fore-grounded sound of the engine in a typical NASCAR trope. Owen says, "You just hold on tight sir," as she, airborne in the backseat, tries to grab him and is flung backward as he accelerates. Again, her crotch is featured prominently as she is pitched from side to side. (We see her empty cup flying around but no coffee). In an underpass area (which again, like the parkade offers a total car space) Owen pulls into an enclave and the soundtrack pauses as they wait for the tail car to pass by. Just as Madonna puts a black-gloved hand on the dash-board, Owen squeals out and catches up to the SUV, winking at the driver. The music changes to Wagner's "March of the Walkure" as the car (actually two cars, since one was destroyed in the filming) flies up over an overpass and through urban space that could be any town USA.[4]

After more quick cuts, we reach the climax of the film and height of Owen's skill. As Owen calls, "Well, we got you here, and in good time, too," he simultaneously skids into a 180-degree turn that lands the car expertly at the kerb and the red carpet and, in an impossibly acrobatic move, reaches back to open the Star's door. Pulled by the centrifugal force of the spin, Madonna flies out of the car to land on her butt onto the red carpet. A crowd of photographers awaits her, and stare in shocked silence. Here is where we see the Star's eyes for the first time, although her grimace makes it clear that it may not be "worth it." The post-ejaculatory hush is relieved by a collective gasp from the crowd of truly ugly men, we see Madonna's face, with a rather grotesque expression, looking down at her crotch, and then, as we are shown her coffee-stained trousers, the music begins again and the cast of male photographers begins to shoot. After the significant line-up of credits, a very brief message from BMW fades on and off the screen: "BMW recommends that you always wear your seatbelt."

The film conveys a high level of artifice. The stunt woman looks nothing like Madonna; the road spaces are non-contiguous and there are various continuity, or editing mistakes, such as no coffee appearing on her crotch until the last shot; the font choice of the title shot "star" plays on the science fiction of Star Trek. Even the casting of Madonna in this role emphasizes the fictional. The Star, the stunt woman, and Madonna all overlap in the representations here: the stunt woman is bruised up, but our pleasure is that Madonna herself is both in on it and "getting it." Despite the semi-slapstick approach and Madonna's obvious approval of the story line, make no mistake: a star is being beaten in this film. The scenes were shot with a stunt woman who had to dye her hair, since they could find no way to make a wig stay on. Guy Ritchie testifies in his director's dub that this professional stunt woman was "pretty bruised up," after the shooting. Six cars were used in the shooting of the film – and traces of the chase are left all over the road in skid marks and scattered vehicles. Much of the film time is taken with the Star getting a good thrashing.[5]

The film plays on a series of stereotypical car-film tropes. First, the chase, which can be read only as a pretext for speed since there is actually no real impetus for pursuit. Second, the trope of the video game is clear in what looks like digitized imagery of up-close tires and the joy-stick/gear shift of the car, as well as the setting of various urban nether-spaces. These combine with the characterization of the diffident European chauffeur (Owen is actually from New Zealand), which echoes both James Bond and Owen's own butler role in Gosford Park. Film reviewer Elvis Mitchell describes Owen's character, with his

> exquisite concave facial structure that . . . could have been engineered by exacting Bavarian designers . . . He's the driver, one of those laconic action samurai who are chauffeur, shrink, bodyguard and some kind of master mechanic . . . his elongated and cruel jaw line, suggests . . . a thug Bond.
>
> (Mitchell 2002)

Owen's confident Commonwealth-bred masculinity contrasts markedly with the Star's hysterical bitchiness. It is, in fact, hard to describe her behavior without resorting to gendered words: nag, bitch, hysterical. By the men, the self proclaimed "super star" is referred to as "this lady," "cunt," "sweetheart," "the bomb," and "marm." Yet when she finally registers her complaint against the latter, the designation of sir, along with the changing color of the traffic light begins the sadistic fantasy of her beating.

An audience will also be familiar – if not critically so – with the woman who seems powerful, and yet ultimately is portrayed as both pawn and victim. The film is utterly politically incorrect and stereotypically so. The plot-summary given on the webpage that offers the download, and that is repeated in many of the on-line and newspaper reviews, is this:

> The driver faces perhaps his most perplexing challenge: Coming face-to-face with a hugely talented and successful rock star. But beneath her beauty lies a problem she always gets what she wants [*sic*]. Guy Ritchie directs Clive Owen and a surprise guest star in a battle of power against power.

If BMW's ad department were suddenly taken over by feminist standpoint theorists, the write-up might look more like this: highly successful and rich woman brings on the jealous ire of her male inferiors. In order to get revenge for her success – which they take personally, they hire a driver to give her, in her own manager's words "everything I pay you for, breakfast, lunch, and dinner." The driver takes that as permission to assault her, through the vessel of the car, and ultimately to humiliate her in front of her fans. These fans/paparazzi are all too ready to consume her humiliation in their own ways. The potential BMW consumer ostensibly identifies with the driver, or the car itself and desires the automobile.

These readings demonstrate how social valuations of space – the private space of the car and the social space of the street are utterly soaked in gender relations and how they are understood. In what follows of this chapter, then, I figure out how urban spaces and how the car, particularly through the chase film, renders these legible.

Speed in the City

Streets offer ostensibly public spaces. They are supported through taxes that everyone pays, and are, in theory at least, accessible to any driver. Though in some senses highly regulated spaces, many unlikely mixes occur: fleets tend to be incompatible, leaving room for dangerous mixings of large and small vehicles; abilities and ages of drivers and pedestrians vary drastically; notions of what constitutes fair or good use of the street differ wildly. Furthermore, as the continued – if limited – availability of publicly funded alternative transit makes clear, mobility is understood in political terms to be, if only in an extremely circumscribed way, a right, or at least, a basic need. But Madonna made clear the hierarchies accorded to modes of transportation in the sarcastic comment to her bodyguard: "Take the bus."

"The pleasure of driving" that BMW uses as its logo derives not from the car's inherent superiority to the bus, but is contingent on symmetries among driver, machine, road, and city. Driving is a practice, its triumph overtly depends on the sociomaterial spaces it traverses. In that sense, as much as the city offers a necessary frame for the representation and apotheosis of the car in this advertisement, driving depends on certain versions of urban space, and the car and driver constitute the city as a particular kind of space. So if the motorist cuts a figure that needs to be regulated and governed in accordance with sociability and overall efficiency, the "car and driver" creates urban space that sets the terms for how space can be used by other, less institutionally and materially privileged, users (Jain 2004).

The conditions under which the value of the luxury BMW is created is underpinned and given its dynamic through the way it enables the consumer *cum* driver to partake in the physical space of the city. Putting this analysis in the context of a film rather than of the city itself allows an understanding of the way in which the representational spaces (the car, the city, and the film) nestle together to give meaning to the characters and their qualities: the bitchy, very wealthy, representationally shifty celebrity; the cutting edge, if working class, skilled driver; and the transgressive, highly engineered, automobile. While these spaces are representational, the goal of the film advertisement is to change the urban space itself – to have more BMWs on the road. Thus, it is worthwhile to more closely examine how these modes of space and the characters they offer rest on regimes of physical violence.

After all, the excesses of the BMW are not only in the latest curves of its side paneling, its leather seats, and bullet-proof glass, but in its engineering and capabilities of speed. There is a very physical component to this luxury item, one that enables the BMW driver to transgress urban American spaces that have become congested, tedious, and banal ironically, but precisely, *because* they were built for automobility (Morse 1990; Sheller and Urry 2003). Thus, if luxury serves as the model for consumption and the means of social hierarchy, it also creates and renders normal the material conditions of everyday life, and in turn makes these unlivable spaces "relivable" through (fantasies of) their very infraction and their technical superiority.[6] The BMW films, through the various car chases, make this evident.[7] This recursive relationship, between the city, luxury, and citizenship is what I want to turn to here.

BMW has particularly cast itself in the rhetoric emerging from the safety research of the last four decades and the engineering it spurred. The company website devoted to the 5 Series model, for example, boasts that:

> Behind every aspect of its safety technology is a rigorous intelligence that constantly works to maximize the safety of the car, its occupants and its environment. High-performance brakes, run-flat tires and innovations such as Adaptic Headlights are elements of its precise active

safety system. Passive safety is boosted by the BMW 5 Series' extraordinarily high-strength occupant cell and its sophisticated Advanced Safety Electronics system. Because true driving pleasure demands more than a feeling of safety.[8]

BMW's safety engineering does not come from a place of benevolent interest in public health. These safety innovations take place in the corporate-consumer circuit and are most useful in the context of the BMW's technological superiority: high speeds require strong headlights and higher-performance brakes and airbags. In turn, this superiority is most useful not in terms of the automobile fleet and its habitation of the road as a public space, but, rather, in the kinds of social logics that lead to a social and material one-upmanship. In bringing together a spatio-temporal order of the journey, *Star* unselfconsciously relies on these relations of automobility, film, and gaming and thus offers one way of understanding how space is consolidated as a valuable asset that can be taken, and, more specifically, that can be taken with the aid of the speeding commodity – the BMW. Thus, the violence of this advertising film is not something to be denounced, for denunciation obviates the ways that the gendered and classed violence provokes the transgressive luxury brand identity of BMW, mutually, with the representations of the city and of celebrity.

In the film, the space of the car is constructed for the viewer in various ways. Owen is seen primarily through the front windshield, in the parts of his body that touch the controls, and occasionally from behind. The Star is seen, when seated, head-on from the front windshield, and from the side windows when in compromising positions. The car itself is viewed from a variety of angles: it has been filmed by another moving vehicle, with cameras attached to it to show, for example, smoke coming off the tires, and from cameras that look to have been placed along the side of the road. A number of spliced shots illustrate Owen's control over the vehicle, braking, stepping on the gas, and so on, in a way that serves to pull together the action and reaction of the car, the road, and the backseat.[9]

This representation is as much about film and its limits as it is about the car; the history of cinema has pivotally shaped the way we understand these registers of motion, time, and space. Film is the ideal medium for representing the possibilities of the journey as a narrative: with its sole mobile eye it isolates, concentrates, and then follows the event of the journey. Karen Beckman notes this parallelism between the roving eye of the camera and the journey of the car in her reading of the film *Crash*. She writes:

The "rules" of this road (or film) are marked by gutters, bollards, and white lines, all of which perforate the unbroken "strip" of the road's surface like sprocket holes, holes that seem complicit with the singular, mechanical, and unidirectional motion of the road.

(Beckman 2003: 107)

68

As she notes, filmmakers such as Jean-Luc Goddard have conversely used driving as a way of exploring the camera's possibility of motion – to investigate the vantage points offered by automobility and how movement is structured.[10] In the book *Crash*, J.G. Ballard (1973) writes of the ongoing relativity of movement among motor vehicles in this way:

> Two airport coaches and a truck overtook us, their revolving wheels almost motionless, as if these vehicles were pieces of strange scenery suspended from the sky. Looking around, I had the impression that all the cars on the highway were stationary, the spinning earth racing beneath them to create an illusion of movement.
>
> (quoted in Beckman 2003: 196)

The camera offers the potential to reiterate a scene generated in the movement of the car for a still spectator. Walter Benjamin's observation that the film offers a manipulation of space in a "spatio-temporal" order "wherein 'fragmented images' are brought together 'according to a new law'"[11] can be employed to better understand how film can offer an analogy for understanding the "takings" of automobility. Like the camera, the automobile can lift a narrative from place and recode a narrative through a new spatio-temporal structuring. In so doing, the film employs various complementary devices; the combination of music, the soundtrack of women's screams and screeching brakes, and quick-paced editing control the passage of this story and of this trip. It is not incidental, for example, that Italian road signs and music annotation bear similar postings. Directives of "rallentando" and "fermata" guide the passage of an event: a violin sonata or a car slipping through an intersection are both events that negotiate time for their expressive content. In *Star*, the Blur 2 song and Wagner combine with the soundtrack and visuals to guide the viewer through the journey. In this way, *Star* offers a total space: the car, the occupants, and the city, the music, and the story each fit with, and make sense within, the timed narratives of the others.

But if the camera has structured understandings of space, computer games have presented spaces of choice. In *Star*, the city and the other cars offer merely a background to BMW and Owen, and the city provides exactly the kind of obstacle course that can allow him to complete the various acts of teaching the lesson, making the journey, and having some fun. This obstacle course not only looks like, but plays like a video game similar to *Grand Theft Auto III*, the best selling computer game of 2002 in which the player is required to drive across town on various errands and points are accrued for "killing" pedestrians. Indeed, through the fast-paced cuts of Owen's feet on the clutch, accelerator, and brake; his hand on the gear-shift/joy-stick; his eyes in the mirrors; and his hand on the door-handle to let her out, Owen could as easily be playing a video game as driving a car – and indeed, the distinct lack of congested freeways in this Los Angeles scene serves as testament to Owen's "secret" knowledge of the city (the same type of knowledge promised by GPS). The film offers us the space of the car and its environ as a complete space, and as one that, while not completely under his control, Owen solely negotiates and that is there for his enjoyment.

Individuated automobile trips also reconstitute social space through their own narratives of the journey: space experienced from the car is not social space, but carries its own register that has as its goal the privatization, or the taking, of units of space over time.[12] If film creates the illusion of movement through the nimble substitution of still images, the car journey requires the progressive collecting and taking of units of space over time.

The moving space offered by the film or the car is very different from that of, say, the intersection, where each of these journeys, vis-à-vis their registers of speed, navigate around and sometimes collide with one another. As Derek Simons (2005), following de Certeau, has noted, the different registers of the journey come together at the intersection not as flow, but as a series of frontiers – pedestrians crossing streets and drivers making right turns and each base their decisions on the experiences that border their own journeys.[13] The intersection is, perhaps, that place where the inherent social co-operation required by mobility is most obvious, and these are precisely the sorts of spaces that are missing from *Star*. Of course, narrative cohesion requires a narrative exclusion, and the recognition of other registers of space would, quite simply, break the storyline. Therefore, other journeys, other spatio-temporal orders, implicitly become the background for the primary journey represented in the film.[14] To put part of the point in its most banal form, in *Star*, driving is not co-operative, it is transgressive. But these transgressive narratives do not dissipate once completed – they are enabled and made concretely permanent streetscapes, intersections, road signs, freeways, skyscrapers, and all the other trappings of car culture. These, because of the potential for so many journeys by so many car owners, become a permanent taking of social space regardless of the actual journeys taken. The road offers an exclusive space even when empty through the *imminent* takings of the automobile.

Star offers a twenty-first-century American urban flâneur, a means of making meaning of a city, "el armpito" in the Star's own words, that has all but been evacuated of possibility for public or civic life. In his book *Non-places*, Marc Augé discusses the bizarre nether-sociality of airports, hotel chains, and freeways (see also Adey and Bevan, this volume, for a critique of Augé on airports). He writes of the subject traversing these non-places:

> What he is confronted with [and as Augé points out, the subject of these spaces *is* masculine], finally, is the image of himself . . . The only face to be seen, the only voice to be heard in the silent dialogue he holds with the landscape-text addressed to him along with others, are his own: the face and voice of a solitude made all the more baffling by the fact that it echoes millions of others. The passenger through non-places retrieves his identity only at Customs, at the tollbooth, at the check-out counter. Meanwhile he obeys the same code as others, receives the same

messages, responds to the same entreaties. The space of non-place creates
neither singular identity nor relations; only solitude and similitude.
(Augé 1995: 103)

But the freeway is unique among these non-places; experienced through the
potentially (but not necessarily) highly personal place of the car, it offers a
unique combination of public, highly regulated, and commodity-driven space.
Advertising is in the business of relentlessly coding our interpretation and ways
of inhabiting this non-place in its unremitting efforts to build meaning into
experience of driving. Much of this effort has been in distinguishing the car as
a place separate from the space it traverses. With its darkened windows, personal
drink holders, and high-quality sound systems, the privatized vessel of the car
is increasingly the medium of the public non-place of the road.[15]

After Heisenberg, any observer also changes what there is to be observed:
in making meaning, in consuming the city as a byproduct of a journey, the
driver further renders it legible only in its own register. The noise it creates
requires soundproof windows, the CO_2 it emits requires the high-quality carbon
filters, and the speed it generates means that even cyclists need the side curtain
airbags boasted by the BMW.[16] So in the move to render the city legible through
luxury automobility and in bypassing the congested freeway, the driver notches
up the most basic tools necessary for citizenship and the assertion of the right
to move. In a different context, Celeste Langan has put a similar point this way:
"Capital-intensive technologies of amplification – not only of speech, but also
of mobility – have so altered social being that even the unimpaired (but also
assisted) body has the character of a disabled subject" (Langan 2001: 468).
Langan draws our attention to the *physicality* of identity discrimination. Car
culture demonstrates this both through the class inequities of mobility (Bullard
and Johnson 1997) as well as the physical threat of automobile injury and death.

To take the point a step further, when that disability is attached to the
body rather than to the environment that structures it, the "coding" of bodies
by the environment becomes obscured. The way the environment enables and
encodes speed for some inhabitants, to the detriment of others, becomes invisible. Second, if one considers Langan's point in terms of violence, rather than
disability, the point emerges as something like this. When capital-intensive technologies such as automobiles are understood as self-reliant feedback loops
(more speed requires better brakes means more airbags), the result is vast differentials in overall safety, and virtually no oversight for the "public" of the street.
This can be seen in the increasing popularity of sport utility vehicles, where a
perceived safety of SUV occupants has resulted in higher fleet fatality rates.[17]

While I have focused here on representation and the urban form, *Star*
demonstrates how an acceptance of certain formations of techno-social identities (gender, class, celebrity) makes automobility a legible good and undergirds
fantasies about the car as a desirable commodity. At the same time as these
identities are stabilized through automobility, they consistently render logical,
invisible, and acceptable class and gender inequities. They also render invisible

the kind of violence, registered in everyday injuries and deaths, that underpins their circulation, making this violence seem like a result of, rather than a con stitutive part of, the always already gendered, raced, and classed culture of auto-mobility. Ultimately, *Star* demonstrates the differential stakes in automobility that are masked by a liberal belief in the body as a site of natural rights. Consumption, and particularly luxury consumption, not only consolidates and displays social hierarchies, but physically extends, in a real political sense, private desires and goals that determine how social spaces can be inhabited in a way that goes far beyond identity politics.

I think *Star* can be used to take this leftist cultural critique of entitlement and urban space a step further. *Star* plays perfectly to its audience. As Neferti Tadiar said about Manila, urban space constitutes a particular social order as a system of representation. The system of representation offered by *Star* in part echoes, and in part challenges, that offered by the urban Los Angeles in which it is set.

The film offers an array of iconic references: Madonna as the pop culture critic of dominant culture; BMW as the symbol of 1990s dot.com wealth and its youthful, exuberant, excess; coffee both in its addictive capacity and playing on the ridiculing drama of a coffee-burned crotch; the Gosford Park butler playing James Bond. To see the film is to know that everyone playing in it is "in" on the particular array of ironic cultural references – it is to tacitly accept that legal complaints about (McDonald's) too-hot coffee are ridiculous, and to enjoy the physical attack on a woman and a critique of a superstar. People laugh because they identify, and identify with, this critique: thus, viewers have a critical language and, indeed, the film is aimed at an audience that knows how to play semiotic games at a high level. Enjoyment of critique is part of the pleasure of the film.

In part because the film offers a semiotic game, and in part because it is aimed toward a certain class of viewers, no one wants to *defend* urban space, or the sociality of the road – in the mode, of, say, the critique above. Rather, the film interpellates viewers to understand the value of urban space only as a means to transgression, which the luxury car enables. In *Star*, building on the history of American urban car-chase iconography, urban space is the launch pad for individuated fantasy. Luxury and its seemingly universal acceptance as a social good, affirms the righteousness of those fantasies.

Thus, *Star* gives us a model to make sense of the "public" space of the street – in essence by representing it as a resource that can be privatized and taken. Luxury consumption forms a new version of public through a collabor-ation of private, collective desires. All kinds of violences play into this, from the ways that identities glom to and are naturalized by cars, to the kinds of spaces built by and for cars, to the individual and tax-base financial burdens of them, to the enormous physical disempowerment of non-drivers and the unthinkable human toll of traffic accidents.[18]

The environment encodes the means of its habitation: in American urban centers, those means include the fantasies of its infraction, which is precisely what BMW offers. Within the fold of their planned homogeneity, parkades, underpasses, and on-ramps also harbor the terms of their own infraction. The underground garage is always already the dangerous place where an evil man jumps out of the back seat of a car or where one plans a meeting with a double agent or FBI informant. Or, with their wide smooth tracks of asphalt they invite bigger living through speeding or skateboarding. These social narratives are as implicit to these spaces as concrete pillars, painted white lines, or the occasional speed-bump. But the "infraction" itself, much as it underpins the pleasures of automobility, remains invisible, shrouded under the cover of a common sense that understands automobility to be normal usage of the street rather than a good that is distributed by urban design and ideologies of consumption and luxury.

Conclusion

Many theorists have convincingly demonstrated that social regimes such as heterosexuality or whiteness depend on a "constitutive outside," or an Other against which normative socialities can be constructed. Bodies marked by race or gender, it is argued, have more difficulty in fulfilling cultural promises of human flourishing. Ann Stoler, for example has demonstrated that the social injury of race is "woven into the weft of the social body, threaded through its fabric" (Stoler 1995: 69). Physical violence, too, is threaded through the fabric of American commodity culture even as it holds out the promise to resolve the hurts of everyday life. Imaginative and material accessibility to these regimes certainly cloves to certain groups along axes of race, class, and gender, as I have tried to demonstrate here. However, they provide a certain distance from the bodies physically marked by race and gender: anyone can buy a BMW and in so doing, partake more or less in a new social regime (at the same time as Driving While Black will qualitatively change one's ability to partake in luxury, as black drivers tend to be more strictly regulated than white drivers). In this way, the terms of address of the film are not so much along the lines of an identity politics, but rather, dictated by the anticipated presence of the luxury consumer, even as to fully partake in this identity of the "luxury consumer" means certain avowals and disavowals of other identities.

Star fashions a journey, in some sense fantastical, and in another simply reiterative of an everyday car trip. Therefore, it is worth our attention to look critically at this journey's components: a star being beaten, an apotheosis of speed and luxury, the derision of spilled coffee. These events and objects circulate and provide the possibilities for each other. While these may seem like trivial issues, it is my hunch that they have a lot to say about how the politics of consumption creates material worlds. In these worlds, a liberal notion of choice that adheres to some "base" material body is the least effective way of understanding how commodities and citizens intermingle and co-constitute. In fact, as I have tried to outline here, commodity violence is intricately tied in with consumer promises and a commodity regime that comes along with the product, a regime that includes both the institutional and material frameworks that distribute goods such as mobility, and frame complaints such as fraud.

For this reason, attempts to reduce road violence – from speeding tickets to hygiene films to driver's education, have uniformly failed and will continue to do so until they search for different ways to understand the sociality of the road. I offer one entry point for such an analysis here.[19]

Notes

1 Acknowledgements: Deborah Cohler and Derek Simons, who invited me to present the work in its early stages to audiences at San Francisco State University and Emily Carr Institute of Art and Design. I have also presented parts of this work at several conferences: the T2M in Eindhoven, organized by Gijs Mom and Clay McShane, the Alternative Mobility Futures conference in Lancaster, organized by Mimi Sheller and John Urry, and the conference on the city at the University of Surrey, organized by Nina Wakeford. Thanks also to a number of friends who have read drafts of the piece with amazing engagement and audacity, including Nicolas Blomely, Kris Cohen, Deb Cohler, Jake Kosek, Samara Marion, Mimi Sheller, Derek Simons, Anne Stott, and Miriam Ticktin. Without these brilliant readers the piece would not have been nearly as much fun to write. Thanks also to Linda Campani, and other friends and colleagues who watched these bizarre films with me and shared in discussion, and thanks to Cynthia Leighton.
2 These days it is assumed that streets are for cars. But any early history of automobility demonstrates that early streets were taken by automobilers with a great deal of resentment by other users.
3 The description of the film here is virtually the same as that in my article "Violent Submission: Gendered Automobility" (Jain, 2005). The argument that follows, however, is completely different.
4 The 1936 film *Master Hands* (available at prelinger.com) represents all aspects of car manufacturing in a style that Rick Prelinger describes as "Capitalist Realism" (as a play on Socialist Realism). The soundtrack features Wagner.
5 Astonishingly, the gender violence is utterly invisible to many audiences to whom I have shown this film. When I asked over 100 of my students in my Car Culture class at Stanford to fill out a short questionnaire on the film, violence or similar concepts were mentioned only twice, students typically found the film funny, entertaining, or occasionally boring. The two possible reasons for this are first, that the styling and caricature hides the violence as violence, or second, that they simply do not see it; they are immune.
6 "The US has 3.95 million miles of roads – 1.1 miles of road per square mile of land. Based on an average lane width of twelve feet, these roads encompass an area of 12 million acres, equal to the combined area of Massachusetts and Maryland" (quoted in Patton 2004). Patton bases his statistics on Office of Highway Policy Information, *Highway Statistics Series 2001 Federal Highway Administration, Oct. 2002, Table HM-10*, http://www.fhwa.dot.gov/ohim/hs01/index.htm. How can it be that in the field of noise, pollution, space-taking, and so on produced by automobility the luxury automobile driver consistently emerges as the cultural hero?
7 Particularly in *Hostage*, where the already incompetent police don't stand a chance in their Ford Crown Royals.
8 http://www.bmw.com/generic/com/en/products/automobiles/showroom/5series/sedan/index.html (accessed November 24, 2003).
9 In one of the best articles on car culture, Kristen Ross writes in *Fast Cars, Clean Bodies*, that the shared technologies of film and automobiles "reinforced each other. Their shared

qualities – movement, image, mechanization, standardization – made movies and cars the key commodity – vehicles of a complete transformation in European consumption patterns and cultural habits" (Ross 1995: 38). She quotes Louis Chevalier, "The pleasure of driving in the city will become, as the city is gradually effaced, the pure and simple pleasure of driving, the automatism of the automobile" (46).

10 Passolini on the other hand did not use cars in his films because they were too capitalist (Linda Campani, personal communication, September, 2003).

11   If industrialization has caused a crisis in perception due to the speeding up of time and the fragmentation of space, film shows the healing potential by slowing down time and, through montage, constructing . . . a spatio-temporal order wherein "fragmented images" are brought together "according to a new law."
(Buck-Morss 1991: 268)

12 Many commentators have written about how the perceptual space of the car has given rise to what Chester Liebs has called an "architecture for speed reading," or an architecture of neon signs and high buildings that can be interpreted while passing a landscape at 60 mph (Liebs 1985).

13 My colleague Jason Patton describes the problem of registers, or frontiers, as one of flow: the sidewalk has breaks when it comes to a road, whereas the road has no such material breaks.

14 Carrying this observation to the quotidian journey, one might note such "road speak" as the saying "anyone going faster than I am is a maniac and slower is a laggard," or in the double meanings of the word "pedestrian." I believe that one of the key reasons for the animus toward the SUV by many car drivers is due to some sort of moral estimation on how they take too much of the "commons."

15 Several people I interviewed for another project on Sports Utility Vehicles mentioned that these vehicles change the sociality of driving – no longer can you see a wave through a smoked window after having let someone in, no longer can you judge behaviors from a driver's body language.

16 Noise would provide one fascinating political economy of the car. Joel Eastman quotes a Chevrolet's general manager, who boasted in the mid-1950s, "We've got the finest door slam in the low-price field, a real big car sound." Or think about this 1920 ad: "whizzing speed . . . unleashed it will roar nose to nose with an express train" (Eastman 1984: 87). While noise, power, and speed were linked early on in automobile advertising, automobile noise has been linked to a decreased ability to concentrate and sleep. This has serious class consequences in thinking about where highways have been built and who lives near them, as Tom Lewis has written in *Divided Highways: Building the Interstate Highways, Transforming American Life* (1997).

17 Collision deaths account for about 58 percent of vehicle deaths each year, and they have increased slightly since 1980. In that year a total of 10,600 traffic collision fatalities was divided as follows: 6,500 car–car, 500 LDT–LDT, 3,600 LDT–car. In 1999, numbers were, respectively, 3,800, 1,800, and 5,200, for a total of 10,800, with the shift in fatalities attributed to LDT collisions reflecting the predominance of "light trucks" and SUVs on US roads (Summers *et al.* 1998, and Joksch 2000).

18 This argument could be taken a step further to think about how these models of luxury require engagements with capitalism. A colleague from the former East Germany tells me that BMW and Mercedes have now become the indicators of prestigious consumption. Before, he says (with no particular nostalgia), everyone would join the waiting list and eventually have a car – and spend their leisure in community pursuits. Now, however, they work longer hours than ever in order to purchase these perceived "goods."

19 A strong thread of the debates around SUVs is that they materially force the occupants to endanger others in order to keep themselves safe. The truth value in these thoughts is arguable (not at all clear that SUVs are overall safer) – but it is true that in a crash that involves an SUV there is a significantly higher likelihood (than in similar crashes

involving sedans) that there will be a fatality, and that the fatality will be an occupant of the smaller vehicle. The irony, then, is that the existence of SUVs on the road has led to a sort of one-upmanship view of car safety.

Works Cited

Augé, M (1995) *Non-places, Introduction to an Anthropology of Supermodernity*, trans. John Howe, London and New York: Verso.

Ballard, J G (1973) *Crash*, London: Cape.

Beckman, K (2003) "Film Falls Apart: *Crash*, Semen, and Pop," *Grey Room*, 12: 94–115.

Buck-Morss, S (1991) *The Dialectics of Seeing: Walter Benjamin and the Arcades Project*, Cambridge, MA: MIT Press.

Bullard, R D and G S Johnson (1997) *Just Transportation: Dismantling Race and Class Barriers to Mobility*, Gabriola Island, BC: New Society Publishers.

Eastman, J (1984) *Styling vs. Safety: The American Automobile Industry and the Development of Automotive Safety, 1900–1966*, Lanham, MD: University Press of America.

Jains, S. (2004) "'Dangerous Instrumentality': The Bystander as Subject in Automobility," *Cultural Anthropology*, 19 (1): 61–94.

—— (2005) "Violent Submission: Gendered Automobility," *Cultural Critique*, 61: 186–214.

—— (2006) *Injury*, Princeton, NJ: Princeton University Press.

Joksch, H C (2000) *Vehicle Design versus Aggressivity*, DOT HS 809 194.

Langan, C (2001) "Mobility Disability," *Public Culture*, 13 (3): 459–483.

Lewis, T (1997) *Divided Highways: Building the Interstate Highways, Transforming American Life*, New York: Viking.

Liebs, C (1985) *Main Street to Miracle Mile: American Roadside Architecture*, Boston, MA: Little, Brown.

Mitchell, E (2002) "BMW Hopes that its Mini-Movies Will Sell Cars," *The New York Times*, June 26, 2002. *Critic's Notebook*, Online. Available http://www.murphsplace.com/owen/articles/critics.html (accessed November 18, 2003).

Morse, M (1990) "An Ontology of Everyday Distraction: The Freeway, the Mall, and Television," in Patricia Mellencamp (ed.) *Logics of Television: Essays in Cultural Criticism*, Bloomington, IN: Indiana University Press, pp. 193–221.

Patton, J (2004) "Transportation Worlds: Designing Infrastructure and Forms of Urban Life," Doctoral Dissertation, Science and Technology Studies, Rensselear Polytechnic Institute.

Ross, K (1995) *Fast Cars, Clean Bodies: Decolonization and the Reordering of French Culture*, Cambridge, MA: MIT Press.

Sheller, M and Urry, J (2003) "Mobile Transformations of Public and Private Life," *Theory, Culture and Society*, 20 (3): 107–125.

Simons, D (2005) "Imagicity: Impressive Technology and the Will to Image in Vancouver's Changing Streetscape, 1920–2004," unpublished Ph.D. dissertation. Vancouver: Simon Fraser University.

Stoler, A L (1995) *Race and the Education of Desire: Foucault's History of Sexuality and the Colonial Order of Things*, Durham: Duke University Press.

Summers, S M, Prasad, A and Hollowell, W T (1998) "NHTSA's Research Program for Vehicle Aggressivity and Fleet Compatibility," NHTSA Paper #249.

Tadiar, N (1993) "Manila's New Metropolitan Form," *Differences*, 5 (3): 154–178.

PART II
Re-configuring Co-presence

CHAPTER FIVE

Bypassing and WAPing: Reconfiguring Timetables for 'Real-time' Mobility

Juliet Jain

> *Some people have a vice for reading Bradshaws. They plan innumerable journeys across country for the fun of linking up impossible connections.*
>
> (Daphne du Maurier, *Rebecca*)

Introduction

Transport academics and practitioners regard travel as a means to an end, i.e. accessing a destination activity, and propose that travel time is 'wasted time'. Certainly those fascinated by public transport timetables or who see the journey as a fun activity in itself would be considered eccentric. Travel information, such as the railway timetable, is constructed as a utilitarian mechanism for managing travel time within the context of accessing scheduled activities distributed across geographic space. 'Real-time' electronic travel information, the focus of this chapter, transfers the fun of planning innumerable journeys and impossible connections from printed paper timetables to a computer that selects choices based on minimizing journey times and waiting.

This chapter explores how the presentation of the timetable has developed historically, and how the ability to deliver personalized travel information to someone on the move connects with emerging mobility practices associated with combined use of mobile technologies and existing corporeal mobility infra-structures such as the British national rail network. Specifically, it is concerned with how the traveller is constructed in the discourses of five British 'stake-holder' organizations that are leading in the development of travel infor-mation. These discourses were generated through semi-structured interviews with the Association of Train Operating Companies (ATOC), Totaljourney.com,

Transport Direct, Transport for London and Kizoom, and in the texts contained in related publicity information.

'Real-time' electronic travel information provides the traveller with up-to-the-minute details of service (train, bus, etc.) availability and running time, digitally displayed through a number of media. The traveller can find out whether a train is delayed or on time, or when the next service to a destination might be. This contrasts against the term 'pre-planning', which describes planning a trip well in advance of travelling. 'Real-time' electronic travel information delivered to the individual through mobile devices empowers public transport in competing with the car. It begins to offer the traveller a perception of flexible and spontaneous travel arrangements, usually associated with independent travel modes (e.g. car, cycle or walking), and it reduces time spent waiting, thus extending the concept of the 'seamless journey' to public transport journeys.

A core aim of UK transport policy is to reproduce the experience of public transport travel around the model of car travel and to develop the concept of the 'seamless journey' in order to encourage modal shift from car to public transport (see DETR 1998; Jain 2004). There is an established understanding that travel information can assist modal shift (see, for example, Hepworth and Ducatel 1992). Lyons *et al.*, in developing 'transport visions for the future', state:

> The increasing volume of information present in our lives suggests that information should have an important role in the future development of transport systems. Individuals make travel choices based on their perceptions of the relative merits of alternative options.
>
> (Lyons *et al.* 2001: 76)

Accessing 'real-time' electronic travel information on the move, through WAP, 3G, PDAs, etc., forms part of a developing trajectory in travel information provision within the UK, both on a commercial front and through the government-sponsored organization 'Transport Direct'.

Organizations involved in this trajectory have capitalized on the wide uptake of mobile devices as providing a unique opportunity to deliver information that is personalized to an individual's needs at that specific moment. However, within the academic literature there is no analysis to date of the interface between public transport infrastructures and emergent practices of mobile phone use, which this chapter seeks to address. Specifically, this chapter considers established understandings of time-space organization in everyday life and the social practices emerging with growth in mobile phone ownership to explore how 'real-time' electronic travel information may place specific time-space expectations on public transport infrastructures.

The first section of this chapter reviews the organization of social practice in time and space, and discourses of speed and time savings in the context of mobility infrastructures. It also reviews recent analysis of mobile phone use and the impact of the mobile phone on time-space practices. The second section considers how the presentation of railway timetables has changed and develops an understanding of how electronic travel information becomes personalized. The third section explores how the developers of 'real-time' electronic travel information understand social practice and how they construct notions of 'bypassing' and 'WAPing'. The implications for public transport providers are discussed in the final section.

Mobility and Synchronizing Copresence

Mobility is situated within dynamic networks of social and technical relationships that produce, maintain and dissolve social practices over time. Copresence, i.e. face-to-face encounters with another person or group of people, or with objects and places, has remained a key social practice. While copresence maintains social ties – and there are many instances of social, legal and practical obligation where copresence with another person, place or artefact is essential (Boden and Molotch 1994; Urry 2000, 2002) – the spatial and temporal organization and distribution of copresence has changed in response to urban planning and design, including the technological opportunities for travelling further and faster, corporeally and with virtual communication media (e.g. telephone, email and internet). Achieving the time-space synchronization of copresence opportunities requires transport and communication infrastructures to afford synchronization.

Copresence compels society into sequenced schedules of time-space synchronization based on the temporal orderings of the calendar and clock time. Daily routines are constituted through taken-for-granted scheduled copresent activities such as school, work, meal times, bed times, watching tv, and so on, which together create complex social rhythms (Zerubavel 1981; Nowotny 1994). This time-space ordering is reminiscent of the linear tourist trajectory explored by Spring (this volume). However, first, schedules often fail or are disrupted, as discussed later in this chapter. Second, the social inequalities embedded in the availability of transport opportunities and the spatial and temporal distribution of many activities can limit the choice of how activities are sequenced or even achieved (Church *et al.* 2000; Hamilton and Jenkins 2000; Graham and Marvin 2000; Schwanen *et al.* 2002).

While clock time is central to synchronizing copresence, its social role structures time values and power relations that impact copresent schedules and the management of transport infrastructures. The clock divides time into equal empty homogeneous units that are then filled with social meaning (Adam 1990). This property of clock time lends itself to economic values based on labour time in the exchange process. Time becomes commodified through paid labour and the production process, and this dominant interpretation of time contextualizes other times such as leisure (Thompson 1967; Adam 1990; Glennie and Thrift 1996). The notion that time should be used productively is historically rooted through the 'protestant work ethic', and the concept of 'time = money' further stresses that non-productive time is wasted time. Even the tourist itinerary is based on a notion of using time productively, thus requiring the guides discussed by Spring (this volume).

Economic values and economic power relations are embedded in organizational practices. The demands of work time often push family or personal time to the margin because these other times are economically non-productive and considered more flexible (Daly 1996; Collinson and Collinson 1997). Waiting for another's time reflects a subordinate position and time value differentials, such as an employee waiting to see the boss or a patient waiting to see a hospital doctor (Zerubavel 1981). Gaining access to someone's time in a face-to-face situation instead of dealing at a distance alters the power dynamic and the timing of responses (Boden and Molotch 1994). Transport and communication infrastructures, therefore, are often constructed as serving this economic time structure to ensure the continuation of economic success and productivity (see, for example, the 1998 White Paper on Transport, DETR 1998), yet transport and ICTs have made a profound contribution to society's concepts of time and timekeeping.

Historically, mobility infrastructures have generated new understandings of scheduling practices, and transformed notions of speed and punctuality in line with clock time and economic values of time. The railway, as the first form of mass transport, took a specific role in embedding standardized clock time and the schedule in everyday social practice. Powerful organizations backing nineteenth-century railway expansion in Britain (beginning in Liverpool, as noted by Adey and Bevan, this volume) aided the transformation from the many 'local' time variations to the national time we have today, i.e. Greenwich Mean Time (GMT) (Schivelbusch 1980; Bartky 1989; Lash and Urry 1994). Such was the power of the railway companies that GMT was often referred to as 'railway time'. The clock's central place in the organization of the station infiltrated popular discourse of social arrangements (e.g. 'meet me under station clock') (Richards and Mackenzie 1986). However, despite the romance of station clocks, the railway was another mechanism that disciplined the masses with clock time, through the railway timetable and the need for punctuality (Thompson 1967; Richards and Mackenzie 1986). Specifically, the Victorian railway timetable subjected society to rigid timekeeping practices, which imposed new social demands on punctuality and timekeeping.

Hence, the published railway timetable emerged as an important device for coordinating connections and time-space synchronization (Richards and Mackenzie 1986; Schivelbusch 1980). The railway timetable remains an essential tool for passengers and railway management.[1] It connects with the other temporal rhythms of social and organizational practices, which are implicit in everyday geographic relations (e.g. commuter services serving established working hours), with the assumption that these destination activities provide a demand for rail travel (Fowkes and Nash 1991). Power relationships between passenger and rail company are implicit in the delivery of timetable expectations that facilitate punctual synchronization with destination activities. Punctual and reliable delivery of the railway timetable is high on British transport policy agenda following rail privatization. The concern is that a lack of punctuality and reliability from the rail companies detracts from rail's ability to engage with policy aims of modal shift from road to rail, especially where travel time is perceived in economic terms (for example, see DETR 1998; SRA 2002). Thus, the attention to meeting the timetable has shifted from passengers learning a skill to rail companies meeting the timetable promise. However, in an analysis of rail passenger complaints post privatization, Lyons and McLay (2000) indicate that some passengers made formal complaints that the train would not wait for them (i.e. the train was punctual).[2]

In comparison to rail travel, the rise of automobility in the twentieth century freed society from the constraints of fixed travel timetables and routes, but not the fear of being late. Specifically, the car could compete on distance and journey times with the train. More importantly, for organizing moments of geographically dispersed copresence, the car offered individual spontaneity, the convenience of time-space flexibility with a seamless journey (i.e. leave anytime to travel to any place without changing modes) (Shove 1998). However, the shaping of urban infrastructures, the distribution of services, and other policies with spatial impacts, have assumed the car is the dominant mode of access, which has further constructed the car as a social necessity for the democratic right of social participation (SceneSusTech 1998; Urry 1999; Sheller and Urry 2000; Jain and Guiver 2001). Godskesen (1999: 57) concludes from her research, 'families who don't have cars often plan their activities more carefully to save time', and those who feel 'time stressed' are more likely to buy a second car. Yet, the road congestion and parking constraints produced by increased car use in limited space are turning the car into a less time-space flexible or reliable option, and therefore increasing the time-space flexibility of public transport has become a political goal (Jain 2004).

Information and communication technologies (ICTs) have wrought another set of time-space opportunities for social practice. In particular, new social practices are emerging with the uptake of mobile technologies, such as laptop computers, mobile phones, PDAs and WiFi, which produce a new layer of mobile electronic connectivity over existing corporeal mobility. Established accounts of the impact of ICTs on everyday mobility consider how the instantaneous exchange of data and 'annihilation' of geographic distance can substitute the need for copresence and corporeal mobility (Graham and Marvin 1996; Hanson 1998; Helling and Mokhtarian 2001). The internet and email have relocated paid work back into the home, often with the promise of flexibility around other home-based activities (e.g. childcare and domestic duties), yet also responding to the demands of international time zones for global services and trading (see Steward 2000; Tietze and Musson 2002). Internet shopping, banking, distance learning and other e-services accessible from the home enable time-space practices configured to individual needs rather than managing collective demand (e.g. attending a class), contributing to the move towards a '24-hour' society (see, for example, Kreitzmann 1999). Access to such opportunities via the internet,

Kenyon argues in this volume, can reduce social exclusion. As well as the internet annihilating the barriers of distant access, Wellman (2001) explores the potential of the internet in assisting community cohesion at a geographically local level, arguing from a case study that 'on-line' neighbours had formed more local social networks than those who were not on-line. Similarly, the internet (email, chat rooms and web pages) has played a strong role in maintaining the Trinidadian diaspora and cultural identity (Miller and Slater 2000). However, telephones too, fixed and mobile, have been a key technology in maintaining social networks and organizing copresence (Licoppe 2004).

In the world of instantaneous exchange and virtual presence, Virilio proposes that 'with real-time technologies, real presence bites the dust' (Virilio 1995: 57). Virilio's analysis of acceleration and speed suggests it has redefined social practice and cultural understandings (Zeitler 1999). Yet, the case for the 'compulsion for proximity', outlined above, suggests that face-to-face or copresent interaction remains essential to maintaining and producing social relationships, especially within the realm of work (Boden and Molotch 1994; Adams 1996; Urry 2000, 2002). Copresence has not been completely replaced by ICTs, rather ICTs configure new layers of connectivity and present new opportunities for managing copresence, where the individual is situated within diverse spatial relations.

Research into social practice and the use of mobile phones (or cell phones) exemplifies the interface between the virtual and the corporeal in a mobile society that directly relates to the analysis of mobile electronic travel information in the following sections. Notably, the mobile telephone has emerged as a key tool for synchronizing copresence while on the move (Peters 2002; Sherry and Salvador 2002; Townsend 2002; Perry *et al.* 2001; Brown and O'Hara 2003; Harkin 2003). The mobile phone enables 'real-time' planning, enabling scheduled commitments to become fluid in time and space, and for small amounts of time to be juggled and utilized as they arise. Essentially, this form of time management also assists in the reduction of any time becoming wasted, as a phone call or text message transforms it back into productive time (Eriksen 2001).

The concept of scheduling activities in 'real-time' rather than 'preplanning' constructs the sense of individual rhythms weaving in and out of existing established time-space structures, such as the mobile worker globe-trotting between corporate offices described by Perry *et al.* (2001). Sherry and Salvador (2002) use the metaphor of 'urban jazz' to describe how the improvised rhythms of individual scheduling with mobile phones harmonize with the background rhythms of the city. The challenge is to 'harmonise among multiple flows of activity and the interplay of planned and improvised action' (Sherry and Salvador 2002: 112). Thus, problems of facilitating time-space coordination between mobile workers identified by Perry *et al.* (2001) indicate a challenge for creating moments of fixity within fluidity. Nodes of connection are as important as flows in constituting the urban (Amin and Thrift 2002).

Where punctuality is embedded in power relationships and the notion that waiting is wasted time, the mobile phone also assists in managing pre-planned schedules in 'real-time' to retrieve lost time caused by unexpected ruptures to plans. In a study of mobile workers, Laurier (2002) illustrates how the mobile phone provides a second layer of connectivity while travelling between locations. The mobile phone connects the mobile worker to the corporate base and to clients in distant regions. When the plan is ruptured, i.e. the car breaks down and the worker has to wait for a replacement car, the mobile phone enables, first, the planned schedule to be rearranged, and, second, the worker to find an activity to turn the wait into productive time.[3]

Yet, mobile phones can also contradict established notions of punctuality. Mobile phones enable the complete subversion of fixed time-space scheduling between friends. Peters (2002) suggests that young people use mobile phones to spontaneously arrange and re-arrange meetings in 'real-time' while drifting through the urban environment.[4] The pre-planned schedule is completely dropped in preference for a fluid and transient existence in 'real-time', but this only works when everyone is 'plugged in' to the same communication system. Moving between 'real-time' scheduled moments of copresence requires mobility infrastructures to support this spontaneity, thus time-space flexible modes (e.g. car or walking, depending on distance) are likely to be favoured, especially at night when there are fewer opportunities to use public transport.

In summary, mobility infrastructures are understood to have enabled social relationships to extend and be maintained across space through faster access. Two hundred years on from the first steam train, there are myriad corporeal and virtual connections supporting social networks and practices. Clock time is a device used to manage the complexity of social arrangements in time and space, but with associated economic values of productivity shapes understandings of punctuality, waiting and speed. Thus 'time = money' is embedded into the coordination of copresence and power relationships. Mobile phones, therefore, reassert the notion of being punctual and reducing wasted time by enabling plans to be managed in 'real-time' in response to the unexpected disruption. Yet, the discussion above also indicates that fluid forms of 'real-time' scheduling are emerging with the more widespread use of mobile phones that revoke notions of punctuality, and this will place new expectations on corporeal mobility infrastructures, as the following two sections will explore.

Reconfiguring the Timetable: Print to Digital Display

The discussion above identified that the railway timetable took a leading role in orienting society around clock time. Following the Hatfield rail crash in October 2000, the national rail timetable was suspended with a number of temporary timetables introduced on a week-by-week basis. It brought home the importance of the timetable, in whatever format, for the traveller in negotiating the complex scheduling of spatially distributed activities, which was emphasized in the following interview at the height of the disruption:

> *Interviewee:*   *I think not only do the passengers need the timetable, I think it's also [what] they measure us against. So, you know, it's almost like the promise when you buy your ticket is, this is your journey from A to B and this is when you will arrive. Lots of people plan their days/lives/visits, whatever, off that piece of information. And certainly we've found in the last few weeks when the timetables haven't been available . . . I think the value of them to customers has come home very strongly when it's not available.*
>
> (ATOC, Nov. 2000)

This section reviews how the presentation of the timetable has responded to new technological opportunities that enable schedules to be reorganized, and waiting time to be reallocated, which was envisaged by Hepworth and Ducatel:

> The provision of flexible information in the home and at work is another area for development. . . . Why not locate a terminal in the kitchen? The commuter might then drink those vital last few drops of morning tea before rushing off to the bus-stop, toast in hand.
>
> (Hepworth and Ducatel 1992: 157)

Specifically, in making public transport appear more accessible and flexible to encourage modal shift, there is a political incentive driving the design and delivery of public transport information, and the government-funded Transport Direct is a key stakeholder that perceives its role as coordinating multiple interests within the public and private sectors. The primary shift in information delivery, which this section tracks, is the tailoring of travel information around the individual while serving multiple needs.

Published travel information in print form was the centrepiece of Victorian rail travel. The Victorian rail companies printed and sold their own route timetables, but it was the 'Bradshaw' that 'remained the oracle' (Richards and Mackenzie 1986: 97). The 'Bradshaw's Railway Guide' was first produced commercially in 1839, and over the years increased the detail and quantity of information, which created a reputation of impenetrability. The frustration with Bradshaw's detail, revealed in popular accounts of its use, shows a less standardized railway timetable than today (see Richards and Mackenzie 1986). The challenge of designing printed travel information is that it needs to answer to everyone's enquiries. Wading through the wealth of information can be a barrier for the individual traveller, unless as du Maurier suggests, it becomes a compelling vice.

Reading and translating a timetable for personal travel needs is an art that has to be learnt, and even today, railway timetables tell the reader how to interpret the format. For the frequent traveller this becomes a tacit mobility practice, whose experience is developed over time by 'doing' mobility. A passenger travelling regularly along the same route (e.g. the commuter) may only look at the timetable when it changes. Yet many journeys follow less regular routes, and while the traveller may have knowledge of how to use the railway, he or she will require information on each new route. Likewise, travellers substituting road travel with rail journeys are less likely to have tacit knowledge of the system or even know how to read a timetable. Therefore, simplifying and personalizing data from its generalized format is considered to reduce barriers to public transport travel:

Interviewee: *Information is for everybody, but specifically everyone has their own personal needs, and a system I think only works eventually if it appears to be able to be personalised even though it can never be personal, it can only be general. So that's the trick.*

(ATOC, Nov. 2000)

The printed timetable continues to appear in many forms, from 'spatially fixed' posters at stations to the 'mobile' pocket size planner. Since privatization, each Train Operating Company (TOC) produces its own route timetables for free with varying formats, but the 'National Rail Timetable', consolidating all routes with the complex symbols and caveats reminiscent of the Bradshaw, continues to be printed commercially. Printed travel information has many levels of detail with an expectation of serving multiple needs. The interviewees emphasized the continued importance of the printed timetable as an interface between the public and trains:

Interviewee: *That one of the most popular forms, or perhaps the most popular form, is still the printed timetable and you know, it's nice about the internet and whatever, but you sort of forget about that at your peril.*

(ATOC, Nov. 2000)

Interviewee: *I can't see the day when there will be no paper timetables, or printed maps.*

(Transport for London, April 2001)

Besides the timetable's mass audience, it is the mobility of the printed timetable that is important to this discussion. The timetable remains the same away from the boundaries of the railway (an 'immutable mobile'), and thus enables journey planning in the office, at home, on the bus, or any other location fixed or mobile, as long as its time period is up-to-date.[5] Hence, a person delayed at a meeting can refer quickly to the timetable to check the next convenient departure, adjusting his or her schedule to shunt colliding time commitments, or, someone delayed on one service can seek alternative connections providing he or she is carrying the relevant timetables. Last minute mobility decisions, therefore, can be made with the printed timetable, but no information can be given there to indicate whether services are performing to the timetable in 'real-time'. Time can still be wasted waiting if the train fails to arrive on time. Nor is the information personalized, and not all routes or connections may be indicated.

Train timetable information within Britain is accessible through a range of other media, besides the printed timetable, which move towards simplification and personalization. First, train timetable information is available over the telephone. Post-privatization ATOC set up the one-number National Rail Enquires Service (NRES) call-centre in 1997. Dialling the NRES number connects the traveller to an operator who uses information accessed from a central computer to produce simplified information specific for the person's individual needs for time-route planning in advance.[6] Initially, a telephone pre-planning service was available from local stations working from the fixed printed timetable, which some argue was a less reliable service:

> Interviewee: *Twenty years ago . . . you would have phoned up the local station and when finally the chap answered he would have struggled and given you an incorrect answer, because the technology wasn't there to provide it.*
>
> (SRA, Jan. 2000)

In accessing travel information over the phone, the passenger has to trust an 'expert' to interpret the electronic timetable information.

The second development brings the traveller into a direct relationship with electronically delivered timetable information. Trust is moved from the inter-mediary human 'expert' to individual skill to integrate the 'expert' technology. In discussing the future of passenger rail information, Lyons and McLay (2000) argue fewer people are connected to the internet than to the telephone, but the trend indicates a move towards increased internet access. Travel information and ticket sales came 'on-line' in the 1990s, with rail businesses developing corporate web pages. Railtrack was obliged under its licensing agreement to provide public access to the 'Great Britain Passenger Railway Timetable', and Railtrack developed a website-based journey planner for national rail enquiries. This service was superseded by the NRES website in 2003. Other commercial on-line services include 'www.thetrainline.com' and 'www.total.journey.com'. These services ask the user variations on time/place of departure and destination and provide 'best' matches to the request. The NRES site links to local station 'departure screens' that provide 'real-time' train information, as well as the 'pre-planning' service.

Initially, the phone and the internet were spatially fixed (e.g. the home, office, phone box). However, the mobile phone, the mobile phone in combination with the wireless internet (e.g. WAP, 3G) and PDAs have liberated the spatial access to travel information, much in the same way as the printed timetable itself.[7] Developments in mobile phone technology take this beyond being able to phone a call centre from any location, to more complex data arrangements as it links with the internet, providing there is a signal for data transfer. Thus, the mobile technologies at the centre of this discussion have transformed the now humble mobile phone with internet access into a locus of information, including travel information.

The access to such information via the mobile phone enables greater personalization of data including 'real-time' travel updates while on the move. While this mobile access redefines the passenger's relationship with the travel system, it also redefines the concept of fixed schedules when travelling. New social practices have emerged with the mobile phone, and the information developers see this technology as empowering choice:

Interviewee: *If you watch people travelling, they always seem to have their mobile phone around somewhere, and I guess to be able to check up what's going on, on your mobile phone, will be something which people will want to do, and feel empowered by.*

(Totaljourney.com, Feb. 2001)

Yet these organizations also have to understand and anticipate how these social practices are changing expectations of mobility infrastructures, ways of travelling and the types of travel information required to empower choice.

Kizoom, a British commercial stakeholder in reconfiguring the railway timetable for new communication technologies, has already made assumptions about the way people travel and the type of travel information they desire. The company is a service provider for the delivery of rail and public transport travel information through WAP and PDAs (see www.kizoom.com). Kizoom's services, for instance, enable a travel diary to be logged in at the beginning of the week and the relevant train times to be displayed for the mobile phone user each day. This personalized timetable then connects with a 'real-time' facility that warns of any disruption to 'pre-planned' services and offers alternative suggestions:

Registered users can supply Kizoom with information about their regular travel arrangements and receive real-time alerts of problems, and sugges-tions of alternative routes. This will also allow registered users to speed up their use of the service – by placing 'home' and 'work' bookmarks, users avoid the need to navigate long complex menus.

(http://www.kizoom.com/press/whohot.html,
accessed 18 April 2001)

The company's publicity exploits the underlying assumption of the eco-nomic value of time to reduce the 'wasted' time of delays and waiting, where small amounts of time can be 'spent' on other activities. However, assump-tions regarding the social practices surrounding the organization of everyday schedules are embedded in the design of Kizoom's services.

As with the conventional internet journey planner, Kizoom's WAP service exploits the need to personalize the timetable and reduce its complexity by reconfiguring its presentation for individual needs. 'Real-time' travel alerts extend the notion of personalization. Transport operators view this technological development as a positive interface between traveller and public transport infrastructures:

> Interviewee: *So the WAP technology should enable high levels of personalization, of services. Which has always been the holy grail for customer service, you know. You can't get much more of a holy grail than personalizing generalized services.*
>
> (ATOC, Nov. 2000)

The rail industry and other information providers have recognized the potential of WAP and similar devices, as these technologies have become more incorporated into the everyday scheduling practices of society. Peters (2002) notes that 75 per cent of the UK adult population has a mobile phone, of which a small percentage have WAP or similar:

> Interviewee: *At the moment I think it's something like 6% of mobile phone users are WAP.*
>
> (Kizoom, April 2002)

Despite this small proportion, significant numbers have signed up to services providing a WAP travel information service. For instance, the two 'stakeholder' interviewees indicated there was significant uptake of the WAP information service offered through what was Railtrack's website:

> Interviewee: *Railtrack put up on its site a WAP enquiry service in August of last year, which has been very successful . . . Since it's been up in August, we've not really promoted it very much, it's just been a little button on the site, we have had one million WAP enquiries to the site. And the last time I looked at the registered users we had 25,000 registered users on that site.*
>
> (Totaljourney.com, Feb. 2001)
>
> Interviewee: *We now serve about 60,000 train enquiries a week which is about 8% of the lookups on the site, Railtrack's site.*
>
> (Kizoom, April 2001)

Since conducting this research Kizoom have extended services with more train operators and transport providers (see www.kizoom.com).

The above account indicates that technology has simplified travel information by tailoring it to individual needs. It is no longer necessary to wade through the modern day equivalent of Bradshaw. Travel information can be accessed via the telephone, internet databases and, more recently, the combination of mobile phone and internet (e.g. WAP, 3G and PDAs). This personalization and access has changed the dynamics of travel information. Computerized databases can be updated more frequently with timetable changes, which is particularly important in relation to infrastructure maintenance. It can also contain to-the-minute (real time) information of train running, and thus indicate service delays. Alternative travel options can then be selected. New mobility practices are emerging with these new forms of travel information delivery and new technologies.

Emergent Mobile Practices

The development of electronically delivered travel information requires some understanding of social practices, the organization of time-space, and time values discussed earlier in the chapter, while making assumptions about the future uptake of technologies and the ways in which these will be used. Thus, an imagined user is often constructed within the technological discourses of the developer that makes assumptions about, for instance, the value of time, or where and when travel information is required. This section considers how two forms of mobility practice dominate the discourses of developments in mobile 'real time' electronic travel information, i.e. bypassing disruption and responding to on-the-hoof decision making, which I have termed 'WAPing'.

Designing an information system for future mobility practices is often an iterative process where technical systems develop over time in response to social practice (see, for instance, Suchman 1987; Akrich 1992). For instance, Kizoom was puzzling over the simple questions of how WAP-delivered travel information would connect with social practice and everyday choices through what Suchman calls 'human-machine communication' (1987). The company focused on current social practices to create imagined future mobility practices:

> *Interviewee:*   ... *so people who talk in terms of a really simple, really smart system that's going to make all the decisions and tell you what you do instead, I'd say right now, forget it – the trick right now is to try and give ways of giving information to people in a way that supports the way they currently think. Or the way they think about travelling.* What we've actually got to do is sort of find out a bit more about how that is. *How do people think about that, alternatives, and the way they get to somewhere.*
> (Kizoom, April 2001, my emphasis)

The Kizoom interviewee also made assumptions about the type of person currently using WAP, while recognizing the challenge of defining the future user, whereas the time-space practices of the imagined user across the stake-holder interviews focused on the 'business user':

> Interviewee: *So I think, right now I think there are two extremes. There are a lot of students and 'tec-savvies', and the young kids who go for the latest phone and know exactly what they're doing with it, at one end. And then you get all the sort of business users with high-end brands and so on, at the other. I don't think it's the sort of middle base users who are really using WAP much at the moment.*
>
> (Kizoom, April 2001)

Understanding the time-space practices of other groups to business travellers (i.e. young people, the retired, and those who might use WAP or similar in the future) is equally important for the future interface between travel information, accessed through this technology, and public transport.

The various organizations involved in developing travel information reflected on understandings of current practices that they had observed and experienced as individuals rather than working to a specific model of the future. The discourses presented by these interviews suggested two forms of mobility practice emerging from mobile electronic real-time travel information with the underlying assumption of business travel or commuting. First, the central discourse was of 'bypassing' journey disruptions. The second set of discourses move towards concepts proposed by early social science studies of mobile phone use, which I have termed 'WAPing'. WAPing is fluid movement in time and space facilitated by real-time travel information that supports real-time scheduling across multiple spaces. Both of these mobility practices have implications for public transport provision and concepts of time-space flexibility and service frequency. Beyond the 'business traveller' other travellers may have time-space ties and expectations that are similar or emerge in different ways.

Bypassing Blockages and Disruption

Public transport, especially trains and buses, suffers from an image of delays, disruption and unreliability. Hepworth and Ducatel (1992) view 'real-time' information as a way of saving time wasted at the bus stop or station waiting for delayed services to benefit the individual, and the Kizoom interviewee also indicated similar sentiments about staying in bed half an hour longer if the train to work is delayed. Where transport systems fail to meet the timetable, the trust relationship between transport network and the individual traveller is negated precisely because other schedules are formed around transport timetable expectations as exemplified by the severe disruption to the national railway timetable following the Hatfield crash in October 2000.

'Real time' travel information that indicates where journey delays are going to occur, combined with information on alternative public transport routes, enables the potential disruption to be bypassed and the saving of 'wasted time' thus restoring 'trust' in the system (see also Graham and Marvin 2000). Lyons and McLay (2000) propose 'real-time' information is only necessary for transport networks that are consistently suffering delays and breakdowns and the concept of information to bypass disruption is central to the discourse of information developers.

The mobile phone presents an opportunity for rescheduling meetings due to delays incurred while travelling, and it is now common to hear rescheduling as a content of rail passengers' mobile phone conversations (see also Murtagh 2002). However, relaying 'real-time' travel information through the WAP phone, or similar device, changes the dynamic from rescheduling others to reconsidering travel choices:

> *Interviewee:* *What would be of value to you, to your mobile phone? And one of the few things that people commonly say that definitely would, would be an update on 'is my train going to be on time?' either in the morning or in the afternoon. You know, the commuter, who actually knew not to go to Liverpool St but to go to Fenchurch St, would see that as a really valuable piece of data, whereas half the thing is getting a mobile phone wouldn't really matter, wait until you go home.*
>
> (ATOC, Nov. 2000)

Thus, mobile real-time travel information alters the dynamics of the urban mobility infrastructures. It opens up mobility choices so that time and space of public transport networks are experienced in new fluid ways (Harkin 2003).

The narrative of the interview extract below illustrates how providing 'real-time' mobility information through mobile devices changes the strategies for negotiating travel around London, and although based on the underground ('the Tube') and bus offers a paradigm for all modes of public transport including train:

> *Interviewee:* *... this morning just as I arrived at North Greenwich Tube station, the announcer was telling us that the Jubilee Line was suffering from delays. Now I had already checked the travel news this morning, because I was leaving early because I wanted to get here and get on with work today and I was leaving, and I'd already checked Travel News and there was no hint of problems. Had I known that [there was a delay], I wouldn't have gone that way this morning. But in the end I had to go a completely different way to normal, and* I had to take that decision instantaneously based on no information, very little information, it was just my knowledge of the Tube network. *OK, I won't wait for the ever-waiting train, because it'll be a nightmare,*

> *I'll go north, under the river that way, and then come the long way on the District Line. Now I knew instantaneously that was an option, but for lots of people those kind of choices are everyday choices and it would be so much better for them if they could look at their mobile and it said, [bus] 472 – six minutes.*
>
> (Transport Direct, April 2002, my emphasis)

This interviewee was able to make decisions based on prior experience of London's transport networks, but emphasizes that not everyone has this type of tacit knowledge. This is like hearing travel news on a motorway and not knowing the local roads, which is where an in-car navigation system would offer a similar paradigm. The interviewee used his knowledge to tactically use the interconnecting travel infrastructures to bypass disruption and argues that mobile 'real-time' travel information can enable others to do the same.

More specifically the service provider Kizoom proposed there was a key relationship between the delivery of 'real-time' travel information through WAP, and bypassing disruption or avoiding delays. From Kizoom's data on numbers of log-ins, the interviewee identified specific dates where log-ins peaked, which correlated with events such as the fuel crisis and service disruption on the London Underground due to industrial action:

> Interviewee: *The strike the other day where what happened was actually they (Transport for London) put up posters on the strike in all the tube stations saying 'strike on, find out information at this web url, this WAP url'. Next day we had 5,000 people looking us up. That shows that 5,000 people in London were prepared to put in a WAP url on their phone.*
>
> (Kizoom, April 2001)

The service informed potential passengers where underground services were operating and at what frequency to open up choices within a disrupted system.

Thus, within the scheduled paradigm of everyday mobility practices, 'real-time' information acts in two very specific ways. First, it opens the choice to avoid delays by selecting an alternative route or mode, but second, it also can provide reassurance either that services are running to timetable or that an alternative is available:

> Interviewee: *I think actually if the system works and the real time information is showing that the bus is on time or the train is on time, then that's reassurance.*
>
> (Transport Direct, April 2001)

'Real-time' travel information through mobile devices, such as WAP, maintains 'temporal security' for public transport travellers, preserves the trust relationship between passengers and train operators, and opens up the potential for bypassing when delays and disruption occur.

WAPing Around

The mobile phone supports a new form of 'real-time' scheduling of copresence that 'real-time' travel information can support. I have termed this on-the-move scheduling in conjunction with 'real-time' travel information gained from WAP, 'WAPing around' (see also Sheller and Urry, this volume, on de-synchronization and 'do-it-yourself' scheduling). This is an emergent social practice, which the interviewees recognized, yet the main rationale for 'real-time' travel information was centred on the ability to avoid delay and save time in the context of 'bypassing'.

Early studies of mobile phone use have observed this fluid 'real-time' movement through urban infrastructures (for example, Peters 2002; Sherry and Salvador 2002; Townsend 2002). Pre-planned meetings can be rescheduled in 'real-time', or individuals can drift between moments of copresence in 'real-time' that is responsive to the availability and access to others. For the majority, the individual schedule connects and disconnects with multiple schedules throughout each day; some pre-planned, and others made in 'real-time'. Thus, Sherry and Salvador (2002) use the metaphor of 'urban jazz' to describe how the mobile phone aids the improvisation of the individual schedule to harmonize with background social rhythms.

The Kizoom interviewee revealed how these other time-space commitments interconnect with mobility and the potential for making decisions on the move, and the interconnection between individual rhythms with other schedules:

> *Interviewee:*   *... one of the issues is, though, that the way we think about stuff is actually very, very subtle and complex and – when you're thinking about something really simple, like your travel decisions about going from A to B, might well be, sort of in the background, there are all sorts of things like well, if I went that way then I could nip into this shop and pick up that thing I've been meaning to pick up for months – or if I went that way, then I could spend half an hour with my friend who I've been wanting to see. Or maybe if I stayed the night then I'd be able to do something or go to an exhibition or something I'd like to do.*
>
> (Kizoom, April 2001)

This flexibility and spontaneity of decision-making is usually associated with individual travel modes (e.g. the car), thus 'real-time' travel information through the mobile device (e.g. WAP) potentially opens up public transport for similar spontaneous decision-making:

> *Interviewee:*   *But the mobile element allows people to get information to people* literally as they're making their journey . . . *You know, circumstances change, and they want to pre-plan, literally now, on the street, without having to go to a phone or terminal or anything, but this 'll get information to them.*
> (Transport for London, April 2001, my emphasis)

Electronic travel information via WAP is able to provide the personalized detail when and where it is required, without the traveller lugging around and searching through collections of complex printed timetables, as with the Bradshaw, or having extensive knowledge of routes.

The time-space flexibility of 'WAPing' is discussed mostly in relation to mobile workplace practices. Mobile technologies such as the laptop computer and the mobile phone have loosened spatial ties to the office, and, combined with new employment practices, facilitate mixtures of home working, 'hot-desking', and working on the move. For example, Laurier's study of the 'mobile office' (2002) and Perry *et al.*'s study of mobile workers (2001) illustrate the practices of mobile employees in scheduling and connecting virtually (mobile phone) and corporeally (by car, train and plane) between locations, although not with WAP. The employee who has become nomadic in time and space, sheds the regular time-space practices of the traditional commuter, and thus, for public transport travel may lack the tacit knowledge of local and national transport systems gained by the commuter following a regular route:

> *Interviewee:*   *I think there's also more uncertainty in working. So you change arrangements far more, say, you know, you either work late or you get diverted on the way to work to a meeting or you get rung up on the way home saying no don't go home, meet me at so-and-so. And I think again, all those things tend to create demand for information.*
> (ATOC, Nov. 2000)

Beyond the work place, another area of flexible scheduling is associated with leisure practices, notably after work, connected to the concept of the development of '24-hour cities' (see Kreitzman 1999; Bromley *et al.* 2003). The mobile phone is the essential accessory for people drifting through the night across the city from one copresent arrangement to another made in 'real-time'.[8] Arguably, this drifting in leisure time-space is reflected in the increase of numbers of mobile phone enquires to NRES (around 20,000 calls per night):

> *Interviewee:*   *A lot of it is ringing for information while people are out. So we've got a big peak of information after 9 at night, that we think largely is people in pubs, whatever, saying 'when's my train home?'. So you know, you get the phone*

call at 10 o'clock, can you tell me the next train to so-and-so, and then another one at 11 o'clock, because they didn't actually leave at 10 o'clock, so they phone again. And that seems to be heavily mobile [phone] dominated.
(ATOC, Nov. 2000)

This mirrors the 'mobile' practices of young people discussed by Peters (2002). Young people are adept at drifting around city environments and scheduling and re-scheduling meetings with friends using mobile phones (Peters 2002). This type of practice can only occur with the mobile phone because it precisely is mobile and enables the use of last minute timeframes (i.e. 'real-time').

WAPing assumes the flexibility of other people's schedules as well as that of the individual concerned. In many instances other schedules are less flexible, and while this is usually framed through the business meeting, other social practices such as childcare, education, hospital appointments, etc. still adhere to rigid timetable expectations. Thus, there are likely to be social differences as to who can 'WAP' around and when. Likewise, the availability of public transport services will affect how these systems can support the fluid 'real-time' scheduling of 'WAPing' around.

Implications of 'Bypassing' and 'WAPing' for Public Transport Providers

Travel providers and policy makers are developing mobile travel information (via WAP and similar mobile devices) to encourage greater use of public transport as an alternative to the car. The car is perceived as flexible and responsive to the construction of individual schedules, whereas rail and bus timetables have to interconnect with other schedules and potentially limit time-space flexibility. Mobile travel information (e.g. WAP) intends to contribute to the removal of barriers to public transport use by personalizing and simplifying data in 'real-time', but this, with other emerging mobile practices, creates expectations of what public transport infrastructures can provide.

The discussion of the concepts of 'bypassing' and 'WAPing' highlighted two groups of emergent mobility practice. First, 'bypassing' considered individuals who need to save time and avoid delay due to other fixed schedules, thus who needed to 'bypass' journey disruption to meet these scheduled arrangements. Without delays and disruption this information would not be required, but it is unlikely in such a complex system as the rail network that delays and disruption can be completely eradicated. Second, the concept of 'WAPing' was used to argue that mobile phones have enabled the development of a scheduling practice in 'real-time' while on the move. Both 'bypassing' and 'WAPing' assume that there are flexible mobility options in place to support these practices as and when required.

Many of the discourses presented in the interviews were based on the travel opportunities in London and the south-east where dense layers of public transport (i.e. underground, bus and national rail) provide the time-space flexibility for bypassing and WAPing. The assumptions for these mobility practices are, first, that public transport networks operate a high frequency of services that loosen the constraints of the timetable, and second, that modal integration is in place to facilitate spatial flexibility. Outside of the south-east or other large cities with dense transport networks, or even outside core hours (e.g. late at night or early in the morning) public transport networks are much more limited in frequency and spatial coverage.

The ability for 'bypassing' or 'WAPing around' is constrained to specific areas by the infrastructure provision for corporeal mobility. In particular, it is difficult to imagine how 'WAPing around' could translate from car to public transport when travelling over greater distances across Britain. Likewise, bypassing disruption on a long-distance rail journey may be constrained by the time taken to navigate other routes. Thus, imaginative solutions to supporting these mobility practices need further consideration.

For rail travel, in particular, 'bypassing' and 'WAPing' also require flexibility of other railway practices such as ticketing. Currently the most flexible tickets are the most expensive, thus excluding these practices from many rail passengers. However, flexible ticketing between modes (e.g. bus and rail) and the virtual ticket bookable through the mobile internet are part of the rail industry's discourses of developments arising from these new mobility practices.

The practices of 'bypassing' and 'WAPing' can be see to be directing new ways of organizing public transport, but at the moment these are in their infancy. The rail industry and travel policy need to consider implications of the interface between social practice, mobile travel information and public transport infrastructures in new and creative ways if an argument is to be made that the value of such travel information can enable the creation of seamless public transport journeys as a substitute for car journeys.

Notes

1   The railway operates two timetables. The first provides details of all train movements (freight and passenger – in and out of service), showing passing points and signal information. The second is a simplified version for the passenger. Staff rosters are also dependent on the timetable, as well as safety.
2   Complaints may be due to other transport connections being late and trains being unable to wait for late connections rather than just passenger inefficiency.
3   The female protagonist of this ethnographic account phones a nearby beauty salon and books in an appointment while she waits.
4   An article run by *Elle* magazine (November 1998) illustrates the city of the night is becoming configured around transient mobility shaped 'real-time' by mobile phone scheduling as people drift between groups of friends distributed in pubs and clubs, and provide a customer-base for shops and gyms opening 24 hours as these facilities are used on the way home.

5   Normally the rail timetable is produced twice a year, but engineering works that force frequent changes often cause the publication of supplementary timetables for different time periods.
6   Subsequently, NRES has introduced 'real-time' travel information of each service.
7   Currently, few WiFi connections are truly mobile, i.e. providing internet access while moving. However, train operating companies are testing WiFi and rolling out the service in 2004/2005. The development of mobile WiFi will impact on the future opportunities for travel information delivery.
8   The *Elle* magazine article on 24-hour cities presented the Nokia 8810 as 'the 24 hour essential to synchronise your socialising' (*Elle*, November 1998: 74).

Works Cited

Adam, B (1990) *Time and Social Theory*, Cambridge: Polity Press.
Adams, J (1996) *Can Technology Save Us?* OCED Conference Paper 'Towards Sustainable Transportation', Vancouver, Canada.
Akrich, M (1992) 'The De-Scription of Technical Objects', in W Bijker and J Law (eds) *Shaping Technology/Building Society: Studies in Sociotechnical Change*, Cambridge, MA, and London: MIT Press.
Amin, A and Thrift, N (2002) *Cities. Reimagining the Urban*. Cambridge: Polity Press.
Bartky, I R (1989) 'The Adoption of Standard Time', *Technology and Culture*, 30: 25–56.
Boden, D and Molotch, H L (1994) 'The Compulsion of Proximity', in D Boden and R Friedland (eds), *NowHere: Space, Time and Modernity*, Berkeley, CA: University of California Press.
Bromley, R, Tallon, A and Thomas, A (2003) 'Disaggregating the Space-Time Layers of City-Centre Activities and their Users', *Environment and Planning A*, 35: 1831–1851.
Brown, B and O'Hara, K (2003) 'Place as a Practical Concern for Mobile Workers', *Environment and Planning A*, 35 (9): 1565–1588.
Church, A, Frost, M and Sullivan, K (2000) 'Transport and Social Exclusion in London', *Transport Policy*, 7: 195–205.
Collinson, D and Collinson, M (1997) '"Delayering Managers": Time-Space Surveillance and its Gendered Effects', *Organisation*, 4 (3): 375–407.
Daly, K (1996) *Families and Time: Keeping Pace in a Harried Culture*, London: Sage.
DETR (1998) *The White Paper on Transport. A New Deal Better for Everyone*, Department of the Environment, Transport and Regions, London.
Elle magazine (1998) '3am Club, 4 am Supermarket, 5 am Gym', November, 73–78.
Eriksen, T (2001) *Tyranny of the Moment. Fast and Slow Time in the Information Age*, London and Sterling, VA: Pluto Press.
Fowkes, A and Nash, C (1991) *Analysing Demand for Rail Travel*, Aldershot: Avebury.
Glennie, P and Thrift, N (1996) 'Reworking E. P. Thompson's "Time, Work-discipline and Industrial Capitalism"', *Time and Society*, 5 (3): 275–299.
Godskesen, M (1999) 'Families and Time Use', in J Beckmann (ed.), *Speed – A Workshop on Space, Time and Mobility*, Copenhagen: The Danish Transport Council.
Graham, S and Marvin, S (1996) *Telecommunications and the City: Electronic Spaces, Urban Places*, London and New York: Routledge.
—— (2000) *Splintering Urbanism: Networked Infrastructures, Technological Mobilities and the Urban Condition*, London and New York: Routledge.
Hamilton, K and Jenkins, L (2000) 'A Gender Audit for Public Transport: A New Policy Tool in the Tackling of Social Exclusion', *Urban Studies*, 37 (10): 215–234.
Hanson, S (1998) 'Off the Road? Reflections on Transportation Geography in the Information Age', *Journal of Transport Geography*, 6 (4): 241–249.
Harkin, J (2003) *Mobilization. The Growing Public Interest in Mobile Technology*, London: Demos.

Helling, A and Mokhtarian, P (2001) 'Worker Telecommunication and Mobility in Transition: Consequences for Planning', *Journal of Planning Literature*, 15: 511–525.

Hepworth, M and Ducatel, K (1992) *Transport in the Information Age: Wheels and Wires*, London and New York: Belhaven Press.

Jain, J (2004) *Networks of the Future: Time, Space and Rail Travel*. Ph.D. thesis, Department of Sociology, Lancaster University.

—— and Guiver, J (2001) 'Turning the Car Inside Out: Transport, Equity and Environment', *Social Policy and Administration*, 35 (5): 569–586.

Kreitzman, L (1999) *The 24 Hour Society*, London: Profile Books.

Lash, S and Urry, J (1994) *Economies of Signs and Space*, London: Sage.

Laurier, E (2002) 'The Region as a Socio-technical Accomplishment of Mobile Workers', in B Brown, N Green and R Harper (eds) *Wireless World – Social and Interactional Aspects of the Mobile Age*, London: Springer Verlag London Ltd.

Leyshon, A and Thrift, N (1997) *Money/Space: Geographies of Monetary Transformation*, London: Routledge.

Licoppe, C (2004) '"Connected Presence": The Emergence of a New Repertoire for Managing Social Relationships in a Changing Communication Technoscape', *Environment and Planning D: Society and Space*, 22: 135–156.

Lyons, G (2003) *Future Mobility – It's About Time*, Universities Transport Study Group Conference, Loughborough, January 2003.

—— and McLay, G (2000) 'The Role of Information in the U.K. Passenger Rail Industry', *Journal of Public Transportation*, 3 (3): 19–41.

——, Marsden, G, Beecroft, M and Chatterjee, K (2001) *Transport Visions: Transport Requirements*, London: Landor Publishing Ltd.

Miller, D and Slater, D (2000) *The Internet: An Ethnographic Approach*, Oxford: Berg.

Murtagh, G (2002) 'Seeing the "Rules": Preliminary Observations of Action, Interaction and Mobile Phone Use', in B Brown, N Green and R Harper (eds) *Wireless World – Social and Interactional Aspects of the Mobile Age*, London: Springer Verlag London Ltd.

Nowotny, H (1994) *Time: The Modern and Postmodern Experience*, Cambridge: Polity Press.

Perry, M, O'Hara, K, Sellen, A, Brown, B and Harper, R (2001) 'Dealing with Mobility: Understanding Access, Anytime, Anywhere', *ACM Transactions on Computer-Human Interaction*, 8: 323–347.

Peters, S (2002) 'Rethinking Networks: Identity and Connectivity in the Mobile Age', Absence Presence: Localities, Globalities and Method Conference, Helsinki, 10–12 April 2002.

Richards, J and Mackenzie, J (1986) *The Railway Station: A Social History*, Oxford: Oxford University Press.

SceneSusTech (1998) *Car-systems in the City. Report No 1*, Department of Sociology, Trinity College, Dublin.

Schivelbusch, W (1980) *The Railway Journey. Trains and Travel in the 19th Century*, Oxford: Blackwell.

Schwanen, T, Dijst, M and Dieleman, F M (2002) 'A Microlevel Analysis of Residential Context and Travel Time', *Environment and Planning A*, 34: 1487–1507.

Sheller, M and Urry, J (2000) 'The City and the Car', *International Journal of Urban and Regional Research*, 24 (4): 737–757.

Sherry, J and Salvador, T (2002) 'Running and Grimacing: The Struggle for Balance in the Mobile Work', in B Brown, N Green and R Harper (eds) *Wireless World – Social and Interactional Aspects of the Mobile Age*, London: Springer Verlag London Ltd.

Shove, E (1998) 'Consuming Automobility', SceneSusTech, Sociology Department, Trinity College, Dublin.

Steward, B (2000) 'Changing Times. The Meaning, Measuring and Use of Time in Teleworking', *Time and Society*, 9 (1): 57–74.

Strategic Rail Authority (2002) *The Strategic Plan*, SRA, London. (Also at http://www.sra.gov.uk.)

Suchman, L (1987) *Plans and Situated Action. The Problem of the Human-Machine Communication*, Cambridge: University of Cambridge Press.

Thompson, E P (1967) 'Time, Work Discipline, and Industrial Capitalism', *Past and Present*, 38: 56–97.

Tietze, S and Musson, G (2002) 'When "Work" Meets "Home"', *Time and Society*, 11 (2/3): 315–334.

Townsend, A (2002) 'Mobile Communications in the Twenty-first Century City', in B Brown, N Green and R Harper (eds) *Wireless World – Social and Interactional Aspects of the Mobile Age*, London: Springer Verlag London Ltd.

Urry, J (1999) 'Automobility, Car Culture and Weightless Travel', SceneSusTech, Sociology Department, Trinity College, Dublin.

—— (2000) *Sociology Beyond Societies: Mobilities for the Twenty First Century*, London and New York: Routledge.

—— (2002) 'Mobility and Proximity', *Sociology*, 36 (2): 255–274.

Virilio, P (1995) *The Art of the Motor*, Minneapolis, MN: University of Minnesota Press.

Wellman, B (2001) 'Physical Space and Cyberspace', in L Keeble and B Loader (eds) *Community Informatics. Shaping Computer-Mediated Social Relations*, London and New York: Routledge.

Zeitler, U (1999) 'Mobility, Time Pollution and Ethics', in J Beckmann (ed.) *Speed – A Workshop on Space, Time and Mobility*, Copenhagen: The Danish Transport Council.

Zerubavel, E (1981) *Hidden Rhythms. Schedules and Calendars in Social Life*, Berkeley, CA: University of California Press.

CHAPTER SIX

Reshaping Patterns of Mobility and Exclusion? The Impact of Virtual Mobility upon Accessibility, Mobility and Social Exclusion

Susan Kenyon

Social Exclusion, Accessibility, Mobility and the Internet

Online activity promises to reshape patterns of social exclusion, by reshaping accessibility in space and time. Virtual mobility, via the Internet, is now emerging as a viable alternative to physical mobility as an accessibility technology, providing access to opportunities, services, social networks and other goods. Internet use can overcome space/time constraints, not only replacing existing travel, but also providing an additional means of access to activities from which people were previously excluded. However, for virtual mobility to reduce social exclusion, it must provide the same function as physical mobility or, at least, fill the accessibility gap that a lack of physical mobility leaves. In addition, it must avoid any negative social or mobility effects, which could worsen mobility-related exclusion, or social exclusion in general. If online activity were to increase the spatial and temporal diversity of offline activities, or to otherwise increase barriers to access, those unable to overcome these barriers may experience even greater social exclusion.

Social exclusion as a concept originated in France in the late 1980s and was adopted into the UK political mainstream following the election of the Labour government in 1997.[1] A shorthand term for a complex phenomenon, social exclusion refers to the process by which individuals and communities experience multiple forms of disadvantage, with multiple causality. The author has developed the following definition of social exclusion:

Social exclusion is a multi-dimensional concept, the process of the interplay of a number of factors, unique to the individual or group, the consequence of which is the denial of access to the opportunity to participate in the economic, political and social life of the community. This process results not only in a diminished material and non-material quality of life, but also in tempered life chances, choices and a reduced level of citizenship.

Following from this definition, when considering what it means to be socially excluded, it can be useful to deconstruct the term into a number of dimensions. The author considers nine dimensions of exclusion: economic, living space, mobility, personal, personal political, organized political, social networks, societal and temporal (Kenyon *et al.* 2002). Thus, potential exclusionary factors within the economic dimension include poverty and unemployment; in the living space dimension, factors within the local environment including the availability of local services and the local ecology; in the mobility dimension, the negative effects of both too little and too much mobility; in the personal dimension, factors including age, disability, ethnicity and gender. The personal political dimension considers control over one's own life; the organized political, influence within formal decision-making bodies. The social networks dimension refers to our relations with others; the societal, to wider social issues including crime, education, healthcare and housing; and the temporal dimension, to the impact of time pressures upon one's ability to participate in the community in which one lives.

In common with the theme of this book, this chapter focuses upon social exclusion in cities. Ambivalence exists regarding conceptualization of 'the city', a debate not entered into here (see e.g. Amin and Graham 1997). However, an element of this debate is relevant here: the concept of city flows, especially in terms of the accessibility or time-space interconnectivity of these flows. Social exclusion can be seen to result from the process of exclusion from these flows, which in turn excludes individuals and groups from activities and, thus, from participation. Clustered[2] exclusion within cities is the focus of many policy initiatives to tackle social exclusion, evidenced in the work and publications of the Social Exclusion Unit (SEU), in particular the work of the Neighbourhood Renewal Unit (NRU) (for example, SEU 1998, 2000, 2001, 2004) and is reflected in the principal measure of social exclusion used by government, the Indices of Deprivation (ODPM 2004). Within this, much of the focus is upon deprived inner city neighbourhoods, which can be pockets of social exclusion, in which many of the potentially exclusionary factors within the dimensions identified above are concentrated. Of particular relevance for this discussion are those in the living space and mobility dimensions, each linked in cause and consequence by issues of accessibility.

Lack of adequate accessibility – here, defined simply as the 'get-at-ability' of a destination (Hillman *et al.* 1973, cited Hine and Grieco 2003) – is a key influence upon, and outcome of, social exclusion.[3] The ability of individuals and communities to access and be accessed is a strong determinant of their ability to participate, economically, politically and socially. However, cultural change, allied with changes in the structure of the built environment, have acted to increase the importance of mobility for accessibility, increasing mobility dependence in society, such that those without adequate mobility can find it difficult to access the activities necessary for participation in the society in which they live (see S. Jain, this volume).[4] As such, socially excluded areas and individuals often suffer from a 'poverty of access' (Farrington *et al.* 2004) – a dearth of services in the locality and both the lack of, and the presence of obstacles to, physical access. With regard to the former, such areas/individuals have low levels of formal services including, in the public sector: benefits and other local government support services; educational facilities, including schools and further/higher education; employment services; and healthcare services including chemists, dentists, family planning, GPs and hospitals. With regard to the private sector, areas/individuals can lack banking and other financial services; employment opportunities; food shops and prepared food outlets selling healthy food; and other retail outlets.[5] Levels of informal service provision, including childcare, community facilities, informal learning, religious groups and social support groups, tend similarly to be low. Poor transport-links then compound accessibility issues, reducing the accessibility of opportunities, services, social networks and other goods that lie outside of the local area. Socially excluded areas tend to be ill-served by public transport, to have low levels of car ownership and to have an inappropriate, poorly maintained or unsafe walking/cycling environment. There is now a wealth of literature exploring the exclusion of individuals from mobility, which is caused (in summary) by the low acceptability, accessibility, affordability and availability of both motorized and non-motorized modes (DETR 2000).[6] In light of this inadequate mobility, individuals can, in turn, find accessibility to be inadequate and thus, in such situations, participation can be denied and social exclusion can occur. Table 6.1 summarizes the impacts of low mobility for social exclusion.

In providing a means of access that is not dependent upon physical mobility, virtual mobility could provide an effective way of tackling these exclusionary dis-benefits of too little mobility, without exacerbating the negative community, environmental and spatial externalities associated with increases in mobility. However, as mentioned above, if virtual mobility is to successfully tackle the mobility-related aspects of social exclusion, it must be able to fill the accessibility gap that a lack of physical mobility leaves; to meet the secondary benefits of mobility; and it must avoid any negative social or mobility effects, which could worsen mobility-related exclusion or social exclusion in general. The remainder of this section considers each of these points in turn.

A number of authors have discussed the ways in which Information and Communications Technologies (ICTs), specifically the Internet, are reshaping concepts of accessibility and interaction in space and time. Drawing upon Hagerstrand's space-time theory of accessibility (1970), in which accessibility (or, as defined above, 'get-at-ability') is conceived in terms of the distance

Table 6.1 *Accessibility, Mobility and Social Exclusion*

Dimension of exclusion	Influence of lack of mobility: examples
1 Economic	Unemployment – inability to take a job because of lack of adequate transport to interview and to place of employment. 38 per cent of jobseekers say that lack of transport is a barrier to employment; 12 per cent say that lack of transport has stopped them from attending interview.
2 Living space	Geographical isolation and low level of service provision locally – lack of mobility reinforcing isolation and difficulty accessing key services. Every year, 1.4 million people miss out on medical help because of transport difficulties.
3 Mobility	The cost, routing, timing, accessibility of public transport and the cost and accessibility of private transport acting as inhibitors to access to opportunities, services, social networks and other goods.
4 Personal	Not directly linked to mobility – factors including ethnicity, culture, gender.
5 Personal political	Personal disempowerment – linked to low levels of knowledge/poor access to information and support networks.
6 Organized political	Low participation – linked to inability to travel to meetings, which are often in the evenings in centralized locations.
7 Social networks	Loneliness and isolation – lack of adequate transport to visit family and friends, or to meet new friends. 18 per cent of non-car owners find it difficult to see family and friends because of transport problems; as do 8 per cent of those with access to a car.
8 Societal	Poor educational opportunities – inability to travel to learning venues. A problem for young and old potential learners. 6 per cent of students have missed college because they cannot afford the transport; 6 per cent of 16–24 year olds have rejected FE because of transport costs.
9 Temporal	Time poverty – time taken to travel reduces time for activities, a problem for private and public transport users alike.

Note: All statistics are taken from SEU 2003. Statistics and examples included are indicative, not exhaustive.

between the individual and the activity destination, alongside the time that it will take to traverse this space, such authors suggest that ICTs can act as a new spatial technology, a viable alternative to physical mobility in allowing the user to overcome spatial and temporal constraints to participation. In their discussion of the impacts of ICTs upon accessibility, Batty and Miller (2000) consider how three principal approaches to measuring accessibility, constraint-based, attraction-based and benefit-based approaches, all centre upon physical distance as the primary factor influencing accessibility. However, when ICTs are used, physical distance is no longer a 'cost', limiting participation. Similarly, Couclelis (2000) builds upon such geo-spatial definitions, considering accessibility to be 'the geographic definition of opportunity', in which 'the opportunity individuals have to participate in necessary or desired activities, or to explore new ones, is contingent upon their ability to reach the right places at the appropriate times'. However, in the information age, Couclelis concedes, place and time become less important and activities become fragmented in space and time. With ICTs, the friction of distance need no longer be a component of measures of accessibility, as spatial and temporal coincidence become less important. Space, time *and thus mobility* need no longer be barriers to participation, as individuals can access necessary and/or desirable activities and/or people, without recourse to physical movement (Couclelis and Getis 2000; Kwan 2000). In this sense, virtual mobility has the potential to make all places equally accessible, both in space and socially, removing the importance of place in the accessibility of, and participation in, cultural, economic, political and social life.

But does virtual mobility provide an alternative to physical mobility? And could virtual mobility overcome aspects of mobility-related exclusion? Returning to the examples given above, ICTs and, more specifically, the Internet, now provide access to many of the services mentioned and could tackle aspects of social exclusion in many of the dimensions, as suggested in Table 6.2. There is also a strong hypothetical role for virtual mobility's accessibility role in supporting the welfare state. A number of commentators have examined the potential role of the Internet in administering health, social care and social support, concluding that the Internet could play a role in providing access to services, including healthcare, education and democratic participation (for example, Burrows *et al.* 2000; Carter and Grieco n.d.; Muncer *et al.* 2000; see also web sites including www.faxyourmp.com; www.mysociety.org). Indeed, the present UK government places high value on access to ICT as a fundamental aspect of citizenship in modern times, with the Internet at the centre of plans to improve the accessibility of public services and a pledge to ensure that all citizens are able to access public Internet facilities by 2005 (Hudson 2003; Selwyn 2002). Thus, virtual mobility could play a role in the delivery of social policy and the reduction of social exclusion, providing access to opportunities, services, social networks and other goods, without increasing and possibly even reducing the mobility burden.

The above suggests that virtual mobility may, in some instances and for some users, take the place of offline participation, in response to a predetermined social need. However, the extent to which this is beneficial for the user will depend upon the quality of this interaction. Concern regarding the social implications of online activities has highlighted disparities in the online and offline experience, in particular the loss of social contact, both during the activity and during the journey to the activity, when mobility is considered to be a social activity in itself. Authors have also considered the possibility of deception and misrepresentation in the online world (including Cornwell and Lundgren 2001; Hamburger and Ben-Artzi 2000); and decreased access to social participation

Table 6.2 *The Influence of Virtual Mobility upon Mobility-related Exclusion*

Dimension of exclusion	Influence of virtual mobility: example	Example web sites (where applicable)
1 Economic	Employment: job vacancies posted; job applications; employment advice and training. Credit: alternative forms of credit and exchange.	jobcentreplus.gov.uk, manpower.co.uk, letslinkuk.org, ebay.com
2 Living space	Overcome lack of local services: education; healthcare information and advice; shopping.	learndirect.org.uk, nhsdirect.nhs.uk, ovacome.org.uk, tesco.com
3 Mobility	Overcoming accessibility deficit resulting from constraints upon mobility through use of the Internet.	–
4 Personal	Overcoming personal limitations: new opportunities for participation; support groups.	rnib.org.uk, gayyouthuk.co.uk
5 Personal political	Empowerment through access to information: single issue support groups; legal information.	prisonersfamilies.org.uk, adviceguide.org.uk, justask.org.uk
6 Organized political	Enhanced democratic participation: political parties; pressure groups; government consultations; contact political representatives; campaign information and support; online voting; online form filling.	labour.org.uk, foe.org.uk, sheffield.gov.uk, faxyourmp.com, inlandrevenue.gov.uk
7 Social networks	Contacting family and friends online: virtual communities of interest; geographically based networked communities; chat rooms/IM; support groups.	well.org, redbricks.org.uk, ubooty.co.uk, ukchat.com
8 Societal	Enhancing the community: community policing; local environmental action; community building.	neighbourhoodwatch.net, bbc.co.uk/crime, ebvonline.org
9 Temporal	Saved travel time: more time for participation in activities.	–

and interaction experienced by those using virtual mobility, highlighting concerns regarding the decline in human relations (loss of conversational skills, decline in face-to-face contact, increased social isolation, loss of community) and the exacerbation of social exclusion for those who are not online – a new form of *virtual mobility*-related exclusion, in light of the reduced accessibility of people unable or unwilling to use ICTs (including Graham 2002; Kraut *et al.* 1998; OII 2003; ONS 2003). Social exclusion could be further exacerbated through the use of virtual mobility if the use of ICTs, specifically, the Internet, leads to increased physical mobility and/or spatial change. A number of (inter-related) mobility effects have been hypothesized. Mokhtarian and Salomon, in various papers, both individually and co-authored, have considered the likely effects of ICTs upon travel, with a specific but not exclusive focus upon tele-working (including Mokhtarian 1990; Mokhtarian and Meenakshisundaram 1999; Mokhtarian and Salomon 2001, 2002; Salomon 1986). In his 1986 paper, Salomon considered three likely mobility effects of ICTs: substitution, enhancement and complementarity. This three-way conceptualization of the interactions between physical and virtual mobility continues to shape the literature today, including work by Black (2001), Golob and Regan (2001), Graham (1998) and Mokhtarian and Salomon, referenced above. In the first scenario, virtual mobility takes the place of physical mobility, as online activities replace their offline counterparts, leading to a reduction in overall travel. If this were the case, virtual mobility could overcome mobility-related aspects of social exclusion, as mentioned above, allowing greater participation in activities for those currently unable to participate because of inadequate physical mobility. However, this is countered by the second scenario, which receives much greater support in the literature: that virtual mobility will lead to greater mobility. Virtual mobility could increase physical mobility in a number of ways. For example, in allowing faster access to and conduct of activities, virtual mobility creates more free time, which may be used for more travel, either to new destinations, or for longer journeys. Second, use of ICTs widens social networks and travel horizons, stimulating the desire to travel. In addition, ICTs may facilitate improvements in the transport network, freeing latent demand and facilitating an increase in mobility. ICTs can also facilitate the conduct of activities during travel, increasing the productivity of travel time and, thus, the tolerance of further travel.[7] Finally, by allowing decentralized conduct of activities, ICTs allow less travel on some days, thus increasing tolerance of distance on others, therefore failing to reduce overall journey times.

If virtual mobility results in increased physical mobility, the overall effect will be to further compound both the accessibility burden and mobility-related exclusion, in two ways: reinforcing the exclusion of those with too little mobility; and compounding the negative effects of too much mobility, which tend to be experienced to a greater degree by poorer communities, particularly those in inner city areas. Increased car use will reinforce the cultural and spatial trends that together have increased car dependence in society, further reducing immediate accessibility and increasing both the number and spatial range of activities in which people must participate in order to be included in the society in which they live (Couclelis 2000; Ross 2000; Shen 2000). Trends in public transport provision, largely in response to the cultural, economic and political valuation of car use, which has led to its promotion and increasing use, further reinforce the inaccessibility of centralized services, for those without unrestricted access to a car (Torrance 1992).[8] In her compelling study, Wajcman (1991) presents evidence of the division of communities by roads and the barriers in the built environment to walking/cycling accessibility, which further decrease accessibility for non-car users (supported by authors including Berman 1982; Cahill 1994; DETR 2000). The negative environmental effects of mobility, including land take, resource depletion and, more notably, pollution, create unappealing living spaces, both ecologically and with regard to anti-social behaviour and crime; and 'hypermobility' (Adams 2000) impacts negatively upon health, increasing asthma, obesity and road traffic accidents, each of which are experienced to a greater extent by poorer communities (Acheson 1998).

Thus, there are very real dangers in promoting virtual mobility, if virtual mobility were to result in greater physical mobility, or were to have negative social effects, as hypothesized in the literature reviewed above. However, there is as yet little evidence to support the above hypotheses. Such evidence as there is draws conclusions from the comparison of disparate sources, primarily cross-sectional, quantitative surveys, not designed for use in this way and which can therefore produce unreliable results (DTLR 2002). Recent research, based upon qualitative evidence, challenges the above theories regarding Internet/ personal travel interactions (Kenyon *et al.* 2003), finding little evidence to support the suggestion that travel time saved as a result of online activity is used for increased travel. It is also suggested that a key effect of virtual mobility is to act as a *substitute* for *additional or future trips*, thus dampening the increase in mobility that would otherwise be necessary to facilitate accessibility and social participation. In addition, the existing literature considers the impacts of virtual mobility across society as a whole, rather than specifically for socially excluded groups. The majority of studies focus upon the mobility effects of teleworking, rather than the range of activities from which people can be excluded, described above. Furthermore, given these assumptions regarding the type of person to be virtually mobile and the type of activity that will be conducted, the literature assumes car ownership on the part of the virtually mobile which, given the levels of car ownership, access and use in the UK (DfT 2004), may not hold true. There is also an assumption regarding a desire to substitute, rather than supplement, offline activities, assuming a limited utility value of online activities vis-à-vis their offline counterparts. Linked to this, the literature assumes a constant travel time budget, a human desire or need to travel for a given time period each day. This economic modelling assumption is quite rightly beginning to receive challenge (Mokhtarian and Chen 2004). Equally, there is evidence to suggest that online activity boosts social contact, widening social networks. In addition to enhancing individuals' ability to overcome accessibility constraints, providing access and enabling participation in activities without recourse to physical mobility, online activity can develop community, both online, as geographically dispersed communities of interest (for example, Baym 1995; Hampton 2002; Rheingold 2000; Wellman 1999) and offline, where community intranets have reinvigorated the local community (including Day and Harris 1997; Hampton 2003; Hampton and Wellman 2003; Shearman 1999). Finally, much of the present literature remains rooted in the present, examining how ICTs can enable the user to accomplish existing offline activities online, rather than considering how activities may evolve, new types of activities that can be conducted online, or new ways of conducting activities. In particular, the literature appears closed to the multitasking that is possible online. In projecting the present into the future, the assumptions that we take with us can mean that we fail to see the future as it could be.

The above discussion highlights the need to understand the extent to which virtual mobility is acting as a spatial technology in providing alternative and/or increased accessibility, for whom and with what mobility and social effects, if these mobility effects are to be harnessed or mitigated against for the economic, environmental and social good. However, the accessibility function of virtual mobility and the allied mobility and social effects are unclear. There is a data gap, which prevents full analysis of these impacts. It is the central contention of this chapter that this data gap is due, at least in part, to a methodological gap: the inability of existing methodologies to collect data that is likely to advance understanding of the impacts of virtual mobility. As such, the author has developed a new methodology, the accessibility diary.[9] The following section presents this methodology, before turning to assess the extent to which the accessibility diary collects relevant and robust data that can advance understanding of the interactions between virtual mobility, accessibility, physical mobility and social exclusion.

The 'Accessibility Diary'

The 'accessibility diary' aims to record all activity participation for seven days, alongside the mode of access – physical or virtual – to these activities, to allow assessment of the influence of physical mobility and virtual mobility upon the individual's patterns of participation in activities.[10] In recording access in terms of both physical travel and telecommunications use, the diary aims to assess the extent to which space and time continue to pose accessibility barriers to participation in activities; and the extent to which virtual mobility is substituting for, or otherwise changing, patterns of physical mobility. The methodology described below is the first, to the author's knowledge, to combine recording, at the individual level, of personal travel, activity participation, time use and communications use in one survey. This approach will allow study of the complex interactions between these factors.[11]

The diary design aims to be *useful*, fulfilling study aims and objectives; and to be *usable*, fulfilling user needs for a survey instrument that is quick and easy to use, this latter point being particularly important in light of the length (days) of the diary completion period and the frequency at which the diary must be repeated.[12] Building usability into the diary, itself, naturally reinforces its utility (usefulness), improving data quality and reducing participant attrition. The design of the accessibility diary used in wave one is shown in Figure 6.1. Users are asked for just six basic pieces of information: what they did, the start and end times of the activity; the participation or presence of others; and what else they were doing at the time. The key to the simplicity of the diary for users, yet the rich complexity of the data gathered, lies in the creation of activity codes, designed for this study and reproduced in Figure 6.2.[13]

Day(s) ....

What did you do? *Please write code one main activity*	Start time	End time	Where did you do it? *E.g. at home; at office; between home and work*	Did anyone else do this with you? *Yes/No*	Was anyone else around at the time? *Yes/No*	What else were you doing? *Please enter code and duration for up to three activities*					
						Code	Dur.	Code	Dur.	Code	Dur.

Figure 6.1 *The Accessibility Diary*

Communicating

C1 Face to face
C2 By telephone (landline)
C3 By mobile telephone
C4 By text, or video messaging
C5 By letter
C6 By fax
C7 By email
C8 In chat room

C0 Other communicating
C01 Other communicating – Internet

Information search

I1 Trivia
I11 Trivia – Internet
I2 Window shopping
I21 Window shopping – Internet
I3 Journey information
I31 Journey information – Internet
I4 Employment information
I41 Employment – Internet
I5 Hobbies
I51 Hobbies – Internet
I6 Medical (inc. NHS Direct)
I61 Medical – Internet
I7 News (includes TV, newspaper)
I71 News – Internet

I0 Other information search
I01 Other information search – Internet

Entertainment/recreation

E1 Resting, relaxing
E2 Reading
E3 Do hobbies
E4 Play sports
E5 Cinema, theatre, watch sport, etc.
E6 Social (pub, club, bingo …)
E7 Watching TV, video, DVD
E8 Listening to music, radio
E9 Travelling for pleasure
E10 Surfing (no specific purpose)
E11 Playing computer games

E0 Other entertainment/recreation
E01 Other entertainment/recreation – Internet

Shopping for:

S1 Groceries (main)
S11 Groceries (main) – Internet
S2 Groceries (top up)
S21 Groceries (top up) – Internet
S3 Clothing
S31 Clothing – Internet
S4 Music
S41 Music – Internet
S5 Journeys (not holidays)
S51 Journeys (not holidays) – Internet

S0 Other shopping
S01 Other shopping – Internet

Formal activities

F1 Paid work
F2 Education
F21 Education – Internet
F3 Voluntary work
F31 Voluntary work – Internet
F4 Religious activity
F41 Religious activity – Internet
F5 Campaigns, civic
F51 Campaigns, civic – Internet

F0 Other formal activities
F01 Other formal activities – Internet

Travel

T1 Driving the car
T2 Travelling in car as passenger
T3 Travelling on bus
T4 Travelling by coach
T5 Travelling on train
T6 Riding motorcycle or similar
T7 Travelling in taxi
T8 Riding bicycle
T9 Walking
T10 Travelling on an aeroplane

T0 Other travel

Household and personal

H1 Sleeping
H2 Personal care
H3 Eating, drinking, inc. preparation
H4 Housework, household maintenance
H5 Childcare
H6 Other caring activities
H7 Running errands (e.g. posting a letter)
H8 Escort (includes school run)
H9 Banking, financial
H9I Banking, financial – Internet
Medical (includes GP, hospital)

H0 Other household and personal

Other/Personal

O1 Other activities
O11 Other activities – Internet

O2 Personal activities
O2I Personal activities – Internet

Figure 6.2 The Activity Codes

The activities listed in this study closely reflect the study aims and enable the collection of a breadth of information in one code. As Figure 6.1 shows, this produces a list of activities quite different to those used in other diary studies. Activities with low perceived importance as individually specified activities are grouped into one overarching activity code and include various forms of personal care, or different sports, or hobbies. Such activities tend to fall into the 'entertainment/recreation' and 'household/personal' categories. However, the activities that are more central to the study, for example because of their online/offline substitutability, or because of their influence upon other activities, are coded in more detail. All forms of communication, information search and shopping are listed separately. Travel is recorded as an activity – 'driving a car', 'travelling by coach' – with a different code for each mode. In addition, the Internet is not recorded as a mode of access and online activities are integrated with their offline counterparts.

Achieving the Aims?

Is the data collected via the accessibility diary likely to advance understanding of the interactions between virtual mobility and the nature and extent of mobility and participation? Will the data enhance understanding of the changing nature of accessibility in the information age?[14] In order to alleviate mobility-related exclusion, virtual mobility must enhance accessibility, in one of two ways: either substituting for physical mobility; or supplementing current mobility, providing additional access to activities from which the individual was previously excluded, because of inadequate mobility. As highlighted above, the majority of literature has focused upon the question of substitution. Uniquely, this study will allow assessment of the extent of both substitution and supplementation. Using longitudinal panel data and recording every activity undertaken online and offline, it will be possible to compare the duration and number of online and travel activities over time and to examine any links between the two, at the aggregate and disaggregate activity levels (i.e. by mode and online activity type). It will also be possible to assess online and travel durations by key characteristics; and to assess duration of participation in key activities and any links between duration, mode or number of travel activities, examining the links between mobility and social exclusion, the latter judged in terms of participation.[15] In addition, the variety of activities undertaken can be assessed and any links between such change and online/travel time exposed. The diary is unable to record precise location of activities, beyond home/work/travelling/other. Thus, changes in the cityscape cannot be assessed. However, time spent travelling can be used as a proxy for distance to indicate activity patterns and could be plotted, such that change in distance to activities could in turn suggest the impact of virtual mobility upon the structure of activity space in the urban environment.

Changes in the duration or number of online and travel activities could suggest substitution effects. However, the depth of the dataset will also allow analysis to move beyond simple substitution, towards the identification of supplementation effects. It will be possible to identify the activities that are replacing travel, if appropriate, to assess the likely contribution of these activities to the reduction of social exclusion. In addition, the data will allow assessment of the inclusionary benefits of secondary activities. The ability to multitask while online is a key advantage of this mode of access, allowing the user, for example, to socialize with friends while childminding, overcoming this barrier to participation. Initial results suggest that multitasking adds two days to the average week. Measuring the contribution of ICTs to this, plus the effects of travel time, alongside analysis by activity and personal characteristics, will allow for assessment of the extent to which ICT-use can supplement accessibility and thus provide an alternative to physical mobility in overcoming mobility-related exclusion.

In addition to the hypothesis regarding what virtual mobility must do to contribute towards a reduction in social exclusion is that regarding what it must not do. As outlined above, virtual mobility must not increase physical mobility, nor have negative social effects. With regard to the mobility effects, preliminary analysis of the first wave dataset indicates no relationship between time spent online and time spent travelling. If this finding is confirmed following more detailed analysis, using longitudinal data, concerns regarding the exacerbation of negative spatial and other externalities may be allayed. Second, with regard to the social effects of virtual mobility, both social activities and the sociability of activities, the latter using the columns 'Did anyone do this with you?' and 'Was anyone else around at the time?', should allow assessment of change and the extent to which virtual mobility facilitates greater or less social contact. The quality of this contact cannot be assessed through the diaries alone and will be explored in parallel qualitative research with the same panel.

Concluding Remarks

In previous research (Kenyon 2003), the author identifies the following relationships between mobility and social exclusion:

1   If a person or area experiences social exclusion, they are likely to have low levels of mobility, because of issues of acceptability, accessibility, affordability and availability of public transport.

2   Low levels of mobility can lead to exclusion, preventing accessibility.

3   Lack of adequate mobility can prevent the success of non-transport social policies that seek to tackle social exclusion, because of the high level of mobility implicit in these policies.

4   High levels of mobility can result in exclusion, isolating the individual from the community and/or creating deprivation by expenditure,

alongside negative environmental and spatial externalities, which can exacerbate social exclusion.

The contribution of both low and high levels of mobility to social exclusion is clear. However, understanding of the possible contribution of virtual mobility, principally via the Internet, is in its infancy. This chapter has presented a new methodology, which aims to assess the implications of virtual mobility for accessibility, physical mobility and social exclusion. The theoretical basis of the research, highlighting the complex interactions between accessibility, physical mobility and social exclusion, has been outlined, alongside the weaknesses of existing studies in this area. The weaknesses of these studies have been attributed, in part, to a 'methodological gap' – an inability of existing methodologies to collect data that can advance knowledge of the accessibility implications of Internet use. The 'accessibility diary' has been designed with the hope of filling this methodological gap. The data gathered will allow assessment of the relationship between physical mobility, virtual mobility and social exclusion. By recording all online, offline and travel activities, the accessibility diary will allow assessment of the links between the number, type and duration of offline activities conducted and the number, type and duration of travel activities, plus the same for time spent online. The activity codes, in particular, ensure the depth of information necessary to assess the impacts of virtual mobility upon activity participation, combining assessment of mode of access. Important influences upon substitutability, including sociability factors and the ability to multitask/trip chain, are also recorded.

While, of course, possessing certain limitations, it is hoped that the accessibility diary represents an important methodological advance, allowing the collection of new forms of data, which should provide an equally important advance in understanding of the extent to which virtual mobility is reshaping patterns of accessibility, mobility and social exclusion.

Notes

1 This chapter considers the UK context. Reference to 'government', unless otherwise stated, is to the Labour government, 1997 onwards. Reference to 'policy', unless otherwise stated, is to policy initiatives implemented by this government.

2 Social exclusion can be experienced at the individual and area levels, which Hine and Grieco (2003) term the 'scattered' and 'clustered' levels of exclusion, respectively.

3 Just as social exclusion is a contested concept, so is accessibility, to the extent that, in their discussion of the concept, Couclelis and Getis (2000) are moved to suggest that it is unlikely that there will ever be a single accepted definition of accessibility, or a single accepted measure. Hagerstrand recognizes three constraints upon accessibility in his space-time accessibility theory, which underlies this research: capability constraints, which are discussed above; coupling constraints, which consider the difficulties of coordinating activities in space and time; and authority constraints, in which the impact of power relations upon participation are considered. Definitional debates are undertaken by authors including Batty and Miller (2000), in which constraint-based, attraction-based and benefit-based definitions of accessibility are explored; by Couclelis (2000), outlining

the diverse spatial, functional and economics perspectives on accessibility; by Hanson (2000), whose definition of accessibility as 'reachability, obtainability [and] attainability' encompasses consideration of the role of individual characteristics, including culture, education and information, in accessibility; and Scott (2000) examines the role of accessibility in urban form, perceiving accessibility as variously an attribute, a process and an outcome. Hine and Grieco (2003) question the extent to which physical accessibility necessarily equates to participation, or a feeling of belonging. Thus, one may have access to the workplace, but without qualifications, one cannot participate in the workforce; one may have access to a social club, but without friends, one would not feel a sense of belonging.

4 On the car culture and corresponding changes to the built environment, in which the growth in car use is seen to 'push the world away', see initially authors including Aird 1972; Freund and Martin 1993; Sachs 1992.

5 There is a wide literature discussing these issues, the key aspects of which are effectively summarized in DETR 2000 and SEU 2003.

6 As with accessibility, there is a wide literature discussing transport and social exclusion, including Church *et al.* 2000; Hine and Mitchell 2001; Kenyon *et al.* 2002; Lucas *et al.* 2001; SEU 2003.

7 On travel time, see Lyons 2003, and J Jain, this volume.

8 In addition, there is a rich literature discussing the inadequacies of public transport, considering the journey from door to door, from a range of different perspectives. On children, see initially Thomas and Thompson 2004, Thomsen 2004; people with disabilities, see Christie *et al.* 2000; on gender, see initially Hamilton and Jenkins 1992, 2000; on older people, Bannister and Bowling 2004, Metz 2003; rural areas, ACRE 2001, Farrington *et al.* 2004; and urban areas, Church *et al.* 2000, Solomon 2001.

9 Full discussion of the EPSRC LINK FIT project from which this methodology is taken is outside the scope of this chapter. At the time of writing, full discussion of results has not been published. The reader is referred to www.transport.uwe.ac.uk/research/projects/internet/index.htm for further information.

10 In this study, the diary is used to collect data from *c.*100 participants in the south-west of England, over four seven-day recording periods.

11 Self-administered diaries, completed in real time in a longitudinal panel survey, have a number of advantages over other methodologies for this area of research. First, the enhanced reliability of data collected contemporaneously, as opposed to that collected in retrospective recall or prospective (e.g. stated preference) surveys (Adler 2003; Corti 2001). Second, the volume of data that can be collected, in a single survey. Third, the ability to draw statistical inferences, allowing quantitative measure of change over time; and fourth, data collected over time from the same panel of participants can allow truer measure of change than that possible in cross-sectional surveys (Duncan 2000; Goodwin 1997; Jarvis and Jenkins 2000). The accessibility diary draws upon activity, communications, time use and travel diaries. It is suggested that neither of these methods provides the level of detail necessary to assess the accessibility impacts of virtual mobility, for which it is necessary to record all online and offline activities, alongside duration, mode, location, sociability and multitasking. This chapter avoids chronicling the many different types of travel, activity, time use and communications diaries that have been used, or their relative benefits. The reader is referred to Kenyon 2004a, for full justification of the methodology presented; and to Axhausen 1995 and Harvey 2003.

12 Extensive research has been undertaken to ensure the usability of the diary. In addition to pre-testing and piloting, telephone interviews with participants, a mini-questionnaire, distributed to each participant with the diary and focus groups with participants were undertaken. Full discussion of methodologies and results are available in Kenyon 2004b.

13 Activity codes are commonly used in time-use diaries and have occasionally been used in activity diaries (for example, Gershuny 2000; Jones *et al.* 1983; Keuleers *et al.* 2002). Study of these activity listings informed the initial list of all of the activities that a

person might undertake during the day. In addition, the author recorded all of the activities that she undertook in a week, combining this with lists compiled by a number of colleagues. Every form of online and offline activity was recorded. From this long list of activities, the coding sheet was constructed, containing 89 activity codes, grouped into 8 categories.

14 Data analysis remains in the early stages and, as such, detailed results are not available at the time of writing.

15 The author recognizes that accessibility and participation do not necessarily mean inclusion. For example, one can be employed, but not feel included in the workplace, socially.

Works Cited

Acheson, D (1998) *Independent Inquiry into Inequalities in Health Report*, London: TSO.

ACRE (2001) *Social Exclusion and Transport: the reality of rural life*, unpublished.

Adams, J (2000) 'Hypermobility', Prospect, available http://www.prospectmagazine.co.uk (accessed 27 October 2005).

Adler, T J (2003) 'Reducing the effects of item non-response in transport surveys', in P Stopher and P Jones, *Transport Survey Quality and Innovation*, Oxford: Elsevier.

Aird, A (1972) *The Automotive Nightmare*, London: Hutchinson & Co.

Amin, A and Graham, S (1997) 'The ordinary city', *Transactions of the Institute of British Geographers*, 22: 411–429.

Axhausen, K W (1995) *Travel Diaries: an annotated catalogue*, Innsbruck: Institut für Strassenbau und Verkehrsplanung, Leopold-Franzens-Univers.

Bannister, D and Bowling, A (2004) 'Quality of life for the elderly: the transport dimension', *Transport Policy*, 11: 105–115.

Batty, M and Miller, H J (2000) 'Representing and visualising physical, virtual and hybrid information spaces', in D G Janelle and D C Hodge (eds) *Information, Place and Cyberspace: issues in accessibility*, New York: Springer-Verlag.

Baym, N K (1995) 'The emergence of community in computer mediated communication', in S Jones (ed.) *Cybersociety: computer mediated communication and community*, London: Sage Publications.

Berman, M (1982) *All That is Solid Melts into Air: the experience of modernity*, London: Verso.

Black, W R (2001) 'An unpopular essay on transportation', *Journal of Transport Geography*, 9: 1–11.

Burrows, R, Loader, B, Nettleton, S, Pleace, N and Muncer, S (2000) 'Virtual Community Care? Social Policy and the Emergence of Computer Mediated Social Support', *Information, Communication and Society*, 3: 1.

Cahill, M (1994) *The New Social Policy*, Oxford: Blackwell Publishers.

Carter, C and Grieco, M (n.d.) 'New deals, no wheels: social exclusion, tele-options and electronic ontology'. Online. Available http://www.geocities.com/margaret_grieco/ working/wheels.html (accessed 23 April 2001).

Christie, I, Batten, L and Knight, J (2000) *Committed to inclusion? The Leonard Cheshire social exclusion report 2000*, London: Leonard Cheshire.

Church, A, Frost, M and Sullivan, K (2000) 'Transport and social exclusion in London', *Transport Policy*, 7: 195–205.

Cornwell, B and Lundgren, D C (2001) 'Love on the Internet: involvement and misrepresentation in romantic relationships in cyberspace vs. realspace', *Computers in Human Behaviour*, 17: 197–211.

Corti, L (2001) 'Using diaries in social research', *Social Research Update*, 2: 1–13.

Couclelis, H (2000) 'From sustainable transportation to sustainable accessibility: can we avoid a new Tragedy of the Commons?', in D G Janelle and D C Hodge (eds) *Information, Place and Cyberspace: issues in accessibility*, New York: Springer-Verlag.

—— and Getis, A (2000) 'Conceptualizing and measuring accessibility within physical and virtual spaces' in D G Janelle and D C Hodge (eds) *Information, Place and Cyberspace: issues in accessibility*, New York: Springer-Verlag.

Day, P and Harris, K (1998) *Down-to-Earth Vision: community based IT initiatives and social inclusion*, London: IBM/CDF

DETR (2000) *Social Exclusion and the Provision and Availability of Public Transport*, London: DETR.

DfT (2004) *National Transport Statistics 2002*, London: DfT.

DTLR (2002) *The Impact of Information and Communications Technologies on Travel and Freight: review and assessment of literature*, London: DTLR.

Duncan, G (2000) 'Using panel studies to understand household behaviour and well-being', in D Rose (ed.) *Researching Social and Economic Change: the uses of household panel studies*, London: Routledge.

Farrington, J, Shaw, J, Leedal, M, Maclean, M, Halden, D, Richardson, T and Bristow, G (2004) *Settlements, Services and Access: the development of policies to promote accessibility in rural areas in Great Britain*, Aberdeen: University of Aberdeen.

Freund, P and Martin, G (1993) *The Ecology of the Automobile*, Montreal: Black Rose Books.

Gershuny, J (2000) *Changing Times: work and leisure in post-industrial society*, Oxford: OUP.

Golob, T F and Regan, A C (2001) 'Impacts of information technology on personal travel and commercial vehicle operations: research challenges and opportunities', *Transportation Research Part C*, 9: 87–121.

Goodwin, P (1997) 'Have panel surveys taught us anything new?', in T F Golob, R Kitamura and L Long (eds) *Panels for Transportation Planning*, Boston, MA: Kluwer Academic Press.

Graham, S (1998) 'The end of geography or the explosion of place? Conceptualising space, place and information technology', *Progress in Human Geography*, 22: 165–185.

—— (2002) 'Bridging urban digital divides? New technologies and urban polarization', *Urban Studies*, 39: 33–56.

Hagerstrand, T (1970) 'What about people in regional science?', *Papers of the Regional Science Association*, 24: 7–21.

Hamburger, Y A and Ben-Artzi, E (2000) 'The relationship between extraversion and neuroticism and the different uses of the Internet', *Computers in Human Behaviour*, 16: 441–449.

Hamilton, K and Jenkins, L (1992) 'Women and transport', in K Buchanan, J Cleary, K Hamilton and J Hanna, *Travel Sickness: the need for a sustainable transport policy for Britain*, London: Lawrence and Wishart.

—— and —— (2000) 'A gender audit for public transport: a new policy tool in the tackling of social exclusion', *Urban Studies*, 37: 1793–1800.

Hampton, K (2002) 'Place-based and IT mediated "Community"', *Planning Theory & Practice*, 3: 228–231.

—— (2003) 'Grieving for a lost network: collective action in a wired suburb', *The Information Society*, 19: 1–13.

—— and Wellman, B (2003) 'Neighboring in Netville: how the internet supports community and social capital in a wired suburb', *City and Community*, 2: 277–311.

Hanson, S (2000) 'Reconceptualizing accessibility', in D G Janelle and D C Hodge (eds) *Information, Place and Cyberspace: issues in accessibility*, New York: Springer-Verlag.

Harvey, A S (2003) 'Time-space diaries: merging traditions', in P Stopher and P Jones, *Transport Survey Quality and Innovation*, Oxford: Elsevier.

Hine, J and Grieco, M (2003) 'Scatters and clusters in time and space: implications for delivering integrated and inclusive transport', *Transport Policy* 10: 293–306.

—— and Mitchell, F (2001) 'Better for everyone? Travel experiences and transport exclusion', *Urban Studies*, 38: 319–332.

Hudson, J (2003) 'E-galitarianism? The information society and New Labour's repositioning of welfare', *Critical Social Policy*, 23: 268–290.

Jarvis, S and Jenkins, S (2000) 'Low-income dynamics in 1990s Britain', in D Rose (ed.) *Researching Social and Economic Change: the uses of household panel studies*, London: Routledge.

Jones, P M, Dix, M C, Clarke, M I and Heggie, I G (1983) *Understanding Travel Behaviour*, Aldershot: Gower Publishing Company Limited.

Kenyon, S (2003) 'Understanding social exclusion and social inclusion', *Municipal Engineer*, 156: ME2: 97–104.

—— (2004a) 'Reshaping patterns of mobility and exclusion? Measuring the impact of virtual mobility upon the nature and extent of participation amongst key social groups: a methodology', paper presented at Alternative Mobility Futures Conference, Lancaster, UK, 9–11 January 2004.

—— (2004b) 'An assessment of the usability of the first wave Accessibility Diary'. Online. Available http://www.transport.uwe.ac.uk/research/projects/internet/internet%20working %20paper%201.pdf (accessed 6 October 2004).

——, Lyons, G and Rafferty, J (2002) 'Transport and social exclusion: investigating the possibility of promoting inclusion through virtual mobility', *Journal of Transport Geography*, 10: 207–219.

——, Rafferty, J and Lyons, G (2003) 'Social exclusion and transport: a role for virtual accessibility in the alleviation of mobility-related social exclusion?' *Journal of Social Policy*, 31: 317–338.

Keuleers, B, Wets, G, Timmermans, H, Arentze, T and Vanhoof, K (2002) 'Stationary and time-varying patterns in activity diary panel data', *Transportation Research Record*, 1807: 9–15.

Kraut, R, Lundmark, V, Patterson, M, Kiesler, S, Mukopadhyay, T and Schleris, W (1998) 'Internet paradox: a social technology that reduces social involvement and psychological well being?', *American Psychologist*, 26: 823–851.

Kwan, M P (2000) 'Human extensibility and individual hybrid-accessibility in space-time: a multi-scale representation using GIS', in D G Janelle and D C Hodge (eds) *Information, Place and Cyberspace: issues in accessibility*, New York: Springer-Verlag.

Lucas, K, Grosvenor, T and Simpson, R (2001) *Transport, the Environment and Social Exclusion*, York: YPS.

Lyons, G (2003) 'Future mobility – it's about time', paper presented at 35th Universities Transport Study Group Conference, Loughborough, UK, 4–6 January 2003.

Metz, D (2003) 'Transport policy for an ageing population', *Transport Reviews*, 23: 375–386.

Mokhtarian, P L (1990) 'A typology of relationships between telecommunications and transportation', *Transportation Research Part A*, 24: 231–242.

—— and Chen, C (2004) 'TTB or not TTB, that is the question: a review and analysis of the empirical literature on travel time (and money) budgets', *Transportation Research A*, 38: 643–675.

—— and Meenakshisundaram, R (1999) 'Beyond tele-substitution: disaggregate longitudinal structural equations modeling of communication impacts', *Transportation Research Part C*, 7: 33–52.

—— and Salomon, I (2001) 'How derived is the demand for travel? Some conceptual and measurement considerations', *Transportation Research A*, 35: 695–719.

—— and —— (2002) 'Emerging travel patterns: do telecommunications make a difference?', in H S Mahmassani (ed.) *In Perpetual Motion: travel behavior research opportunities and application challenges*, London: Pergamon.

Muncer, S, Burrows, R, Pleace, N, Loader, B and Nettleton, S (2000) 'Births, deaths, sex and marriage . . . but very few presents? A case study of social support in cyberspace', *Critical Public Health*, 10: 1–18.

ODPM (2004) *The English Indices of Deprivation*, London: ODPM. Available http://www. odpm.gov.uk/stellent/groups/odpm_urbanpolicy/documents/page/odpm_urbpol_029534. pdf (accessed 20 July 2004).

OII (2003) 'Results'. Online. Available http://users.ox.ac.uk/~oxis/enough.htm (accessed 5 December 2003).

ONS (2003) 'Internet access'. Online. Available http://www.statistics.gov.uk/pdfdir/int0903. pdf (accessed 5 December 2003).

Rheingold, H (2000) *The Virtual Community: homesteading on the electronic frontier*, Boston, MA: MIT.

Ross, W (2000) 'Mobility and accessibility: the yin & yang of planning', *World Transport Policy and Practice*, 6: 13–19.

Sachs, W (1992) *For Love of the Automobile*. Berkeley, CA and Oxford: University of California Press.

Salomon, I (1986) 'Telecommunications and travel relationships: a review', *Transportation Research Part A*, 20: 223–238.

Scott, L M (2000) 'Evaluating intra-metropolitan accessibility in the information age: operational issues, objectives and implementation', in D G Janelle and D C Hodge (eds) *Information, Place and Cyberspace: issues in accessibility*, New York: Springer-Verlag.

Selwyn, N (2002) ' "E-stablishing" an Inclusive Society? Technology, Social Exclusion and UK Government Policy Making', *Journal of Social Policy*, 31: 1–20.

SEU (1998) *Bringing Britain Together: a national strategy for neighbourhood renewal*, London: TSO.

—— (2000) *National Strategy for Neighbourhood Renewal: framework for consultation*, London: SEU.

—— (2001) *A New Commitment to Neighbourhood Renewal*, London: SEU.

—— (2003) *Making the Connections: final report on transport and social exclusion*, London: SEU.

—— (2004) *Tackling Social Exclusion: taking stock and looking to the future. Emerging findings*, London: SEU.

Shearman, C (1999) *Local Connections: making the net work for neighbourhood renewal*, London: Communities Online.

Shen, Q (2000) 'Transportation, telecommunications and the changing geography of opportunity', in D G Janelle and D C Hodge (eds) *Information, Place and Cyberspace: issues in accessibility*, New York: Springer-Verlag.

Solomon, J (2001) 'Social exclusion and the provision and availability of public transport', *Transition*, January: 3–8.

Thomas, G and Thompsen, G (2004) *A Child's Place: why environment matters to children*, London: Demos.

Thompson, T U (2004) 'Children – automobility's immobilized others?', *Transport Reviews*, 24: 515–532.

Torrance, H (1992) 'Transport for all: equal opportunities in transport policy', in J Cleary, K Hamilton, J Hanna and J Roberts (eds) *Travel Sickness: the need for sustainable transport policy for Britain*, London: Lawrence Wishart.

Wajcman, J (1991) *Feminism Confronts Technology*, Cambridge: Polity Press.

Wellman, B (1999) *Networks in the Global Village: life in contemporary communities*, Oxford: Westview Press.

CHAPTER SEVEN

Twin Towers and Amoy Gardens:[1]
Mobilities, Risks and Choices

Stephen Little

Introduction

This chapter looks at two events that, by replacing predictability with uncertainty, increased the perceived risk of travel and changed collective understandings of the relationship between place, transport and accessibility. The events demonstrated the vulnerability of core urban systems to threats previously regarded as problems of the periphery.

Just as the Japanese attack on Pearl Harbor represented a pivotal point in US foreign relations and ultimately the emergence of a Pacific Century, so the attack on the World Trade Center towers already represents a pivotal point in the relationship of the US to the international community. The events of 9/11 shifted the paradigm of hijacking which had become almost institutionalized in North America. The expectation, established in the 1960s and 1970s, that political hijackers would seize aircraft in order to make specific demands had been overturned by the time the fourth aircraft hijacked on 9/11 crashed during an attempt by passengers to regain control. The attacks triggered an immediate shut-down of US aviation, which had consequences for both US inter-urban travel and for the global civil aviation system.

The disruption of international travel and trade caused by the SARS outbreak in 2003 was equally damaging for the global economy and impacted equally on transport and communications across East Asia. The rapid diffusion of SARS across and beyond East Asia led to a significant impact on investor confidence, tourism and business travel and was equally disruptive in economic terms. As with the disruption of commerce and trade in New York, companies operating in Hong Kong suspended operations and the Hong Kong and Shanghai Bank (HSBC) implemented a disaster recovery operation with a backup trading floor at an alternative location.

This chapter examines the implications of these traumatic events for the key issues for transport policy and individual decision-making. The continuing 'war on terror' holds significant implications for longer distance travel and international mobility; however, heightened perception of the risks of terrorism and disease produced responses of relevance to both long-distance and local transport.

The proposition that the price of mobility is eternal surveillance is examined through a consideration of risk perception, the mechanisms of recovery and response, and their consequences.

Reframing Mobility

The consequences of 9/11 continue, with varying levels of alert and conflicting demands for security innovations ranging from armed sky marshals on flights to the deployment of new biometric technologies for the identification of passengers.

While these initiatives may or may not be counterproductive, an association between global terrorism and hyper-mobility was forged by the destruction of the twin towers of the New York World Trade Center. Terrorists commandeered the civil transport infrastructure as a counter to the remote application of American power through cruise missile strikes on Afghanistan and Sudan. This subversion of a dominant technology by the marginalized had been foreshadowed by Johnston (2000) in his discussion of 'blowback' from clandestine foreign policy interventions and by Lindqvist (2001) in his analysis of the development and application of military aviation in the twentieth century.

The 9/11 attacks created a new understanding of the nature of co-presence and the dual use of technology. In the context of the unreflective application of technical superiority, particularly air-power, these acts represent a violent repositioning of concepts usually associated with the issues of the global divide and the information divide and the notion of appropriate and appropriated technologies (cf. Little *et al.* 2001). Terrorists unexpectedly appropriated technologies of air transport to bridge the global divide and appropriated the global media to broadcast a gruesome message with the world's undivided attention.

Innovations in information and communication technologies facilitated the current mode of globalization. Over shorter distances transportation and communication technologies were critical determinants of the Western form of urban development and its variants within the global economy (Banham 1960, 1971; Mazlish 1965). The suburban railroad and tramway and the telephone and elevator allowed a combination of high-density business districts and lower density residential settlement.

This model of urbanism became the mode of colonial and post-colonial urban development. In Hong Kong, for example, almost uniformly high-density development is supported by a world-class public transport system, but density also creates problems for the prevention of disease transmission. The 2003 outbreak of SARS (Severe Acute Respiratory Syndrome) was centred on four 35-storey high-rise blocks of flats in a district comprising 19 in total. These contained 350 cases of SARS and were subject to quarantine. It was the vertical transportation within the blocks – the elevators – that turned out to be the key to the spread of infection. Local authorities delivered food and other supplies to the buildings, but the police eventually discovered that half of the occupants had fled. The governments of both China and the Hong Kong Special Administrative Region subsequently implemented quarantine camps for SARS patients.

Before the emergence of SARS a range of transport and communication vectors had already been identified in the spread of infectious disease:

- As an economy measure, airlines reduce the rate of cabin air change in long-distance aircraft. This has been cited both as a vector for diseases and a contributing factor in passenger misbehaviour in the form of 'air rage' (Andersen 1999; Sahiar 1994).

- In addition to long-haul aviation, both long- and short-distance passenger rail transport in the US have been identified as the vector for multiply drug-resistant forms of tuberculosis originating in Siberia (Boseley 2004; Reichman and Tanne 2001).

- Military grade anthrax has been distributed via the US mail, one reason for initial speculation that SARS might be a form of bio-terrorism.

Both new and old, now multiply resistant, communicable diseases will have as profound an impact on risk perception and choices for communication and transport in this century as terrorism. While some argue that risk is a defining characteristic of the current period (Beck 1992), in a broader historical context, we are returning to a more normal situation in which risk and mobility are intimately related.

Govers and Go (2004) suggest that the response to the threat of SARS was disproportionate to the actual risk. The post-9/11 down-turn in air travel and the use of substitute forms of communication evident in the US was driven as much by the potential disruption likely in the event of a second attack as by concern for direct individual danger. However, the rapid distribution of SARS via air transportation and the draconian measures taken to control it at locations remote from its origin – in particular the impact on Toronto and the Canadian economy – ensured a significant response from urban populations in affected regions.

Medical prophylaxis was as essential as the gun-boat to the extension of Western influence at the height of European colonial expansion, particularly in sub-Saharan Africa (Headrick 1981). Without this protection, mortality rates ranging from 50 to 70 per cent in West Africa confined Western presence to coastal trading settlements. Conversely, Western diseases to which indigenous populations had no resistance were critical in the undermining of existing cultures and patterns of trade and settlement in advance of direct contact with Western intruders (Reynolds 1981).

Currently, the mobility of international labour as a component of the low wage service sectors of major urban centres brings fears of both disease and violence from the periphery to the core. For Western travellers, both the overcrowded short-distance metro and the fully booked long-distance flight now hold risks not associated with mobility for many years. These shifts in understanding of exposure to risk and disease can be examined usefully in terms first formulated by Perrow (1984).

Mobility, Risk and Complexity

Perrow introduces the concept of dread and attempts to improve upon cost-benefit analysis and its underlying economic rationality by considering cognitive and social rationalities too.

Perrow is concerned to account for the difference in risk perception between the professionals responsible for the formal accounting of the risk inherent in large-scale high-technology projects and that of the general population exposed to that risk. He argues that people place some importance on the detail or manner of a fatality or accident, and do not confine their evaluation to the measurable outcomes of absolute rationality, that is the simple statistical probability of outcomes. He argues that these differences in formal and informal evaluation reflect a lay understanding of 'disaster potential'.

Perrow draws on work reported by Pfund (1984) comparing the risk perceptions of expert risk assessors and a non-expert group. This shows that though a particular risk, such as nuclear power, might present few or no casualties in a given year, when non-expert test subjects were asked to think about a 'bad year', the respondents revealed an understanding of the scale of possible casualties in situations prone to systems accidents.

The test subjects were also able to rate the different activities in terms of criteria of voluntary participation, scientific understanding, familiarity, lethality, etc. very closely to the experts. However, the experts seemed not to regard characteristics such as 'involuntary, delayed, unknown, uncontrollable, unfamiliar, catastrophic, dreaded and fatal' (the rating of nuclear power) as relevant to the issue of riskiness of the activity, while for lay observers these were significant considerations.

Perrow presents the results of a larger follow-up study of 90 hazards and 18 risk characteristics. Using factor analysis, three principal factors were identified: the most important was a 'dread factor' which was discovered to be associated with:

• lack of control over activity;

• fatal consequences of some sort of mishap;

- high catastrophic potential;

- reactions of dread;

- inequitable distribution of risks and benefits (including the transfer of risks to future generations);

- the belief that risk is increasing and not easily reducible.

The second factor was labelled 'unknown risk' by the researchers and involved risks that are:

- unknown;

- unobservable;

- new;

- delayed in their manifestation.

Examples given in 1984 were DNA research, food irradiation and nuclear power. Clearly a current listing would include Deep Vein Thrombosis and SARS, plus potential bio-terrorism.

The third cluster consisted of risks characterized by 'societal and personal exposure' such as motor vehicle accidents, smoking and pesticides, in which there is some degree of control over exposure.

Perrow points out that all these criteria are characteristics of the complex and tightly coupled systems which are the focus of his book (see Sheller and Urry, this volume). Perrow identifies two dimensions: complexity and coupling. Linear systems are easier to manage than those that are complex and have un-anticipated relationships between components. Loose coupling allows slack to deal with problems; tightly coupled systems immediately propagate the conse-quences of a mishap, and may turn it into a catastrophe. For Perrow, systems that combine tight coupling with complex interactions are problematic. Features of the coupling and complexity described by Perrow are present in the urban regions of developed countries, for example, in the interaction between struc-tural adjustment and the decline in public health standards in the developing countries that are delivering labour for the urban infrastructure.

The narrow economic structural adjustment imposed by Western agencies on developing economies has reversed colonial and post-colonial public health gains (Bello 1999). Migration patterns have delivered the consequences to the developed world. Resistant strains of tuberculosis originating in the far eastern provinces of the former Soviet Union where health services, particularly in the prison system, have effectively collapsed, had already been identified in New York and New Jersey. Corresponding reductions in the social wage have undermined health care delivery in developed economies so that the incomplete application of antibiotic drugs has increased the risk of established diseases through the creation of multiply drug-resistant variants (NIC 2000). TB treatment is now delivered as directly observed therapy (DOT), to overcome the problems created by unsupervised antibiotic treatments.

Other disease risks are emerging with global warming altering the distribution and range of a number of plant, animal and human diseases.[2] Speculation that pandemics may result from the release from polar ice of microorganisms that have not been in the atmosphere for hundreds of thousands of years take these concerns into the realms of Perrow's dread factor.

Bringing it All Back Home: Global Mobilities, Local Risks

At the local level perceived risks from terrorism and from infectious diseases influence decisions about the use of urban mass public transport. At a global level the movement of populations in response to economic and social pressure and the consequent potential for the propagation of both disease and deliberate disruption is expected to continue to increase over the next decade (NIC 2000, 2001). In assessing the risks to the US from global migration up to 2015 the National Intelligence Committee identifies terrorist and organized crime groups exploiting co-ethnic migrant flows and weak migration control.

Globalization as currently understood (Ohmae 1995; Dicken 2003) proceeded from the mid-point of the twentieth century on the back of new information and communication technologies. Nineteenth-century technologies had delivered regularity and predictability through the key technologies of steamship and electric telegraph (Hirst and Thompson 1996). Twentieth-century technologies produced near instantaneous communication, which created new forms of adjacency that offered to replace or supplement physical presence for many purposes.

Large-scale population movements and migration are driven by inequities in development. While tele-working has allowed the outsourcing of jobs from developed to developing locations, a counter-flow continues as evidenced by those casualties of 9/11 who were engaged in key low-wage activities within the World Trade Center. This unevenness of development within and between economies threatens the achievement of sustainability as defined by writers such as Welford (1995) who cites the Bruntland report from the World Commission on Environment and Development (Bruntland 1987).

Unevenness in development is reflected in the interpenetration of the developed centre and the periphery, in terms both of the coordination of dispersed activities through new communication technologies and the physical co-location of high- and low-value activities. The end of the Cold War allowed rapid acceleration of global economic integration. Disparate national and regional cultures are increasingly interacting within networked and globalized economic systems and organizations. Ohmae (1995) refers to the removal of the 'bi-polar discipline' of the Cold War which had obscured differences within and between members of the Eastern and Western blocs and consigned the remainder of humanity to the disparagingly termed 'Third World'. Delamaide (1994) explores the synergies flowing from the re-assertion of historical cultural and economic linkages, offering an alternative understanding to Ohmae's 'zebra strategies' (Ohmae 1995). Ohmae argues that governments should play to the relative strength of the most developed components of national economies in order to create regional synergies. Delamaide suggests that the emergent super-region of Europe represents the reactivation of much older pre-existing geopolitical relationships. The last minute restrictions on the free movement of new citizens of the 25-member European Union is, to some degree, an acknowledgement of both dynamics.

From both perspectives, however, differentials in development are entrenched through dependence upon a global infrastructure constructed around the priorities of the dominant developed economies and the resulting inequities undermine the legitimacy of some national states. As a consequence, globalization and deregulation of economies is producing nomadic communities. These are emerging in response to a complex process of layering of labour markets, both internal and external to the developed economies driving this process. Attali (1991) predicts the emergence of a nomadic international elite, in line with the examples provided by Webber (1964), but movement is not restricted to the elite employees of trans-national corporations. A range of skilled, semi-skilled and unskilled workers, legal and illegal are moving into and between both rural and urban areas of the more developed economies in growing numbers (Castles and Miller 1993).

Migration patterns and improved physical and electronic communications have produced transcontinental extended families in all types of society, and the anxiety and confusion between categories such as asylum and economic migration point to the tensions produced by the growing scale of physical movement within the globalizing system. Remittances from these workers to their relatives and dependants in the home country have become a significant component of global financial flows and they represent a very different form of global workforce from that posited in the mid-twentieth century by writers such as Webber (1964).

Workforce availability and cost in developed economies has been moderated by a degree of tolerance of illegal movement, which has become institutionalized to a degree. According to Kling *et al.* (1991) areas of Orange County, California, the quintessential Reaganite environment, are no-go areas to the INS (Immigration and Naturalisation Service). As economic migration has become conflated with asylum seeking, with terrorism, the permeability of US borders has become a key concern.

Monitoring Mobility

By the end of the first post-Cold War decade it had become clear that a number of institutions were reassessing their roles, just as enforcement institutions shifted their attention from alcohol to other drugs after the repeal of prohibition laws in the US (Grinspoon 1994). The US National Security Agency (NSA)[3] became involved in the development of data verification and encryption during the 1990s, to the extent of proposing standards for commercial transactions that would enable them to monitor traffic. Information networks are emerging as the social milieu of non-place communities. Policy veterans from the Cold War are viewing this non-space arena of global communications as their new fiefdom.

The NSA proposals were vigorously opposed by groups such as Computer Professionals for Social Responsibility (CPSR).[4] However, in the post-9/11 environment calls for the electronic surveillance of either targeted groups, or entire civil populations have gained greater legitimacy. Electronic identity cards, which had been promoted in the late 1980s (Clarke 1989), have again become the solution to the myriad problems of security and control. The need to identify SARS carriers has led to the deployment of infra-red scanning technologies to screen travellers for raised body temperatures. Developments in biometric technologies mean that iris recognition technology is being advanced for both technical reasons – greater accuracy than fingerprints – and cultural reasons – it only requires observation of the eyes, an advantage where individuals adhere to strict religious dress codes (see Adey and Bevan, and Wood and Graham, this volume, on developments in airport security technologies).

Significantly, the SARS outbreak triggered a co-ordinated global response. Both the Center for Disease Control (CDC) in Atlanta[5] and the World Health Organisation[6] provide on-line information on the progress of SARS. An overview of the threat and progress of SARS can be seen at the Globalchange site,[7] which makes a comparison between the responses of front-line health workers to the SARS threat to that of the front-line rescue workers on 9/11.

The US Department of Defense maintains a Global Emerging Infections Surveillance and Response system[8] and APEC (Asia Pacific Economic Cooperation) provides an Emerging Infection network,[9] which includes on-line courses. A world SARS map was established by maptell.com[10] and the measures put in place by various national governments were also published on the World Wide Web. The space, organization and management of the continuing response to SARS can be monitored from any point (Little and Grieco, in press). As a consequence, the reappearance of SARS was met with a prompt and transparent response.

In keeping with the globalized, distributed response, grid computing[11] was being utilized in the analysis of SARS data, as with the SETI at Home distributed screen saver[12] which harnesses the spare capacity of networked PCs in the search for extra-terrestrial intelligence.

Security is also premised on information tracking, control and meta-governance and, the most high-profile information tracking and control arrangements are found in the War on Terror. The National Commission on Terrorist Attacks on the US maintains a web site[13] that presents evidence and findings on-line, and both the CIA view[14] and the FBI view[15] are readily available.

However, these are countered by expressions of concern over the widening of definitions of 'terrorism'.[16] The emergence of the al Jazeera news network[17] has led to the provision of an English-language version of their web site. This provides a contrast to the perspective and the images delivered by Western media. It represents a legitimate and selective use of dominant infrastructures in order to provide a voice for the excluded and demonstrates a reversibility of surveillance relationships.

Recovery, Responses and Consequences

Terror and infection have become new sources of uncertainty in travel and communication and both are intertwined in the prospect of bio-terrorism. These threats are both inherent in travel, but also delivered through the movement of 'others'. The propagation of terror and disease both reflect global mobilities facilitated by ICTs, the enabling technologies of a globalizing economy.

A military definition of security has been perpetuated from the Cold War environment in which surveillance is combined with military intervention in order to achieve some form of stability. As a result of the civil–military communication problems on 9/11, the US civil Air Traffic control system is now mirrored in real time at the air defence headquarters in Colorado Springs. The imperial origins of transport and communication infrastructures are described by Headrick (1981) and the military origins of the Internet have been well rehearsed, and are the key to its robustness. However, tensions between military and civil paradigms are implicit in the less obvious surveillance aspects of this infrastructure. The potential for a surveillance society through 'dataveillance' was identified by Clarke (1989).

The civil paradigm draws on the fruits of 1980s' projects in artificial intelligence and 'knowledge extraction' in the form of social network analysis (Carley and Gasser 1999).

The military paradigm involves high tech weaponry and information warfare, ranging from electronic countermeasures to broadcast propaganda and, ultimately, cruise missile diplomacy. It remains a top-down approach built on the assumption of the ultimate superiority of high-level abstract data. The US is promoting a 'network centric defense' system among its allies. This complex and comprehensive military infrastructure depends upon a highly sophisticated and tightly integrated communication, command and control system. This, in turn, requires expensive specialized equipment constructed to standards dictated by the Pentagon. The UK is committing a significant proportion of its defence budget to achieve compatibility (Dawes 2003). This decision is already impacting as 'rationalization' of front line units is proposed while these same units are engaged in the deteriorating occupation of Iraq (Norton-Taylor and Watt 2004).

Regime change in Iraq was a policy that pre-dated the 9/11 attacks, and was integral to the Project for an American Century[18] in which the US was the only 'superpower'. However, the post-9/11 'war on terror' was the justification for the invasion of both Afghanistan and Iraq. Nation states, whether regarded as ideologically intransigent, corrupt or failing, are poor approximations of an amorphous and mobile enemy.

Despite the US-led Western occupation of both Afghanistan and Iraq, terrorist attacks have continued within Muslim countries. The targets were mobile representatives of Western interests, whether tourists among the Indonesian minority Hindu population in Bali or business travellers around Western banks and consulates in Turkey, a key secular Muslim country.

Tourist travellers had already been targeted by a number of similar campaigns: from the separatist Kurdish movement within Turkey, ETA in pursuit of the cause of Basque separatism in Spain and the IRA seeking economic disruption among the seaside resorts of mainland Britain. In the post-9/11 era, the tourist function of the major urban regions of the developed world is seen as another vector of vulnerability through which risk in the form of terror or disease can penetrate the core.

Surveillance is being promoted as necessary to the management of permissive mobility in the twenty-first century. The UK has the largest population of electronically tagged offenders in the European Community – in excess of 10,000 – and is seeking the development of national identity cards, smart cards and ID cards.

Transport innovations such as the smart cards pioneered in Hong Kong as Octopus and subsequently introduced in London as Oyster give opportunity for further monitoring in addition to credit card tracking and other forms of monitoring electronic transactions. A GPS locator function has been a requirement for all US cell phones since before 9/11. This allows tracking of the user to within a few metres. This facility can be used to deliver location-specific services in high-density environments, or for search and rescue in low-density environments. It also permits extremely fine-grained real-time surveillance of the movement of selected individuals.

Conclusion: Another Route to Secure Mobilities

The collective global tracking of SARS offers a view of more forms of international co-operation than those driven by the war on terror. New international relationships have emerged as the potential of the key technologies have become recognized. China is joining the European Galileo GPS system which offers an alternative to the US system and which is central to both mobile commerce and to security (*People's Daily* 2003). This is an acknowledgement of the potential of a new form of digital divide in which the most developed surveillance systems – those of the US and Britain – conduct real-time monitoring of global Internet connectivity patterns creating a new definition of information rich and poor.

However, many target populations exist with the more pressing problem of 'information asymmetries' formulated by Lamberton (1995). This raises a wider problem than the 'digital divide'. The former concept encompasses the significant proportion of humanity living in communities more than 40-minutes travelling time from even basic telephonic communication: 'information justice' is required (Lamberton 1995), yet through current patterns of migration these populations may have only one or two degrees of separation from the urban heartlands.

The ubiquity of the base technology of the Internet means that access to non-place community does not depend on large investment, nor on esoteric technical skills. Otherwise justifiable criticisms of technological optimism often miss this. ICTs already support the social and economic functioning of diasporic communities, which bridge the global divides (Little *et al.* 2000, 2001).

The components of a bottom-up and networked response to the issues raised in this chapter are already in place. This new paradigm could provide the feedback loop of systems theory and cybernetics (Beer 1981), which would facilitate the development of an inclusive urbanism linking the core and the periphery. This would counter risk and build security through inclusion and engagement rather than deploying the same technology in the vain pursuit of security through an exclusion no longer possible in a globally networked urban form.

Notes

1 Amoy Gardens is the 35-storey high-rise block of flats in Hong Kong in which some 230 people were quarantined for 10 days during the 2003 SARS outbreak.
2 The World Health Organisation suggests that 'a temperature rise of only 1–2 °C over the next 50 years could extend the range of malarial mosquitoes further north – increasing the proportion of the world's population at risk of malaria and other mosquito-borne diseases such as dengue and lymphatic filariasis' (http://www.who.int/infectious-disease-report/pages/ch9text.html (accessed 1 December 2004)).
3 See http://www.nsa.org/ for the Agency's view of 'preparing for the future' (accessed 1 December 2004).
4 See http://www.cpsr.net/ (accessed 1 December 2004).
5 See http://www.cdc.gov/ncidod/sars/ (accessed 1 December 2004).
6 See http://www.who.int/csr/sars/en/ (accessed 1 December 2004).
7 See http://www.globalchange.com/sars.htm (accessed 1 December 2004).
8 See http://www.geis.fhp.osd.mil/ (accessed 1 December 2004).
9 See http://depts.washington.edu/apecein/ (accessed 1 December 2004).
10 See http://www.maptell.com/maps/webmap/world/worldsars.htm (accessed 1 December 2004).
11 See http://www.wired.com/news/medtech/0,1286,58678,00.html (accessed 1 December 2004).
12 See http://setiathome.ssl.berkeley.edu/ (accessed 1 December 2004).
13 See http://www.9-11commission.gov/archive/index.htm (accessed 1 December 2004).
14 See http://www.cia.gov/terrorism/ (accessed 1 December 2004).
15 See http://www.fbi.gov/terrorinfo/counterrorism/waronterrorhome.htm (accessed 1 December 2004).
16 See http://sf.indymedia.org/news/2002/01/114450.php (accessed 1 December 2004).
17 See http://english.aljazeera.net/ (accessed 1 December 2004).
18 See http://www.newamericancentury.org (accessed 1 December 2004).

Works Cited

Andersen, N H (1999) *Broken Wings: A Flight Attendant's Journey*, Vancouver: Avia.
Attalli, J (1991) *Millennium: Winners and Losers in the Coming World Order*, New York: Times Books, Random House.
Banham, R (1960) *Theory and Design in the First Machine Age*, London: Architectural Press.
—— (1971) *Los Angeles, the Architecture of Four Ecologies*, London: Architectural Press.
Beck, U (1992) *Risk Society, Towards a New Modernity*, trans. M Ritter, London: Sage Publications.
Beer, S (1981) *Brain of the Firm*, 2nd edn, Chichester: Wiley.
Bello, W (1999) *Dark Victor: The United States, structural adjustment and global poverty* (2nd edn), (with S Cunningham and W Rau), Amsterdam/London: TNI/Pluto Press.
Boseley, S (2004) 'Deadly strain of TB spread by air travel', *The Guardian*, 16 March, p. 8.
Bruntland, G (ed.) (1987) *Our Common Future: The World Commission on Environment and Development*, Oxford: Oxford University Press.
Carley, K M and Gasser, L (1999) 'Computational Organization Theory', in G Weiss (ed.) *Distributed Artificial Intelligence*, Cambridge, MA: MIT Press.
Castles, S and Miller, M J (1993) *The Age of Migration: international population movements in the modern world*, London: Macmillan.
Clarke, R (1989) 'Information technology and dataveillance', *Communications of the ACM*, 31 (5) May: 498–512.
Dawes, A (2003) 'New weapons for a new doctrine', *Air International*, 64 (5) May: 16–19.
Delamaide, D (1994) *The New Super-regions of Europe*, New York: Penguin.
Dicken, P (2003) *Global Shift: transforming the world's economy*, 4th edn, London: Sage.
Govers, R and Go, F M (2004) 'Cultural identities constructed, imagined and experienced: a 3-gap tourism destination image model', *Tourism*, 52 (2): 165–182.
Grinspoon, L (1994) *Marijuana Reconsidered*, Berkeley, CA: Group West.
Headrick, D R (1981) *The Tools of Empire: technology and European imperialism in the nineteenth century*, Oxford: Oxford University Press.
Hirst, P and Thompson, G (1996) *Globalization in Question*, Cambridge: Polity Press.
Johnston, C (2000) *Blowback: the costs and consequences of American empire*, New York: Henry Holt.
Kling, R, Olin, S and Poster, M (eds) (1991) *Post-suburban California: the transformation of postwar Orange County, California*, Berkeley, CA: University of California Press.
Lamberton, D (1995) 'Communications', in P Troy (ed.) *Technological Change and Urban Development*, Sydney: Federation Press.
Lindqvist, S (2001) *A History of Bombing*, New York: New Press.
Little, S and Grieco, M (in press) 'Electronic Stepping Stones: a mosaic metaphor for the production and re-distribution of communicative skill in an electronic mode', in S Clegg and M Kornberger (eds) *Space, Organization and Management*, Stockholm: Liber.
——, Holmes, L and Grieco, M (2000) 'Island histories, open cultures?: the electronic transformation of adjacency', *Southern African Business Review*, 4 (2): 21–25.
——, —— and —— (2001) 'Calling up culture: information spaces and information flows as the virtual dynamics of inclusion and exclusion', *Information Technology & People*, 14 (4): 353–367.
Mazlish, B (ed.) (1965) *The Railroad and the Space Program: an exploration in historical analogy*, Cambridge, MA: MIT Press.
NIC (2000) *The Global Infectious Disease Threat and its Implications for the United States*, NIE 99–17D, January 2000, Washington: National Intelligence Council.
—— (2001) *Growing Global Migration and its Implications for the United States*, NIE 2001 02D, March 2001, Washington: National Intelligence Council.
Norton-Taylor, R and Watt, N (2004) 'Army chiefs furious at political meddling in proposal to merge Scottish regiments', *The Guardian*, 11 November 2004.

Ohmae, K (1995) *The Borderless World: power and strategy in the interlinked economy*, New York: Harper Business.

People's Daily (2003) 'EU to co-operate with China on Galileo program', *People's Daily*, 28 October 2003.

Perrow, C (1984) *Normal Accidents: living with high-risk technologies*, New York: Basic Books.

Pfund, N (1984) 'Recombinant DNA, miracles and menace', in D Sutton (ed.) *Do No Harm: health risks and public choices*, Berkeley, CA: University of California Press.

Reichman, I B and Tanne, J H (2001) *Timebomb: the global epidemic of multi-drug-resistant tuberculosis*, New York: McGraw-Hill.

Reynolds, H (1981) *The Other Side of the Frontier*, Melbourne: Penguin.

Sahiar, F (1994) 'Economy Class Syndrome', *Aviation, Space, and Environmental Medicine*, October.

Webber, M (1964) 'The urban place and the non-place urban realm', in M M Webber, J W Dyckman, D L Foley, A Z Gutenberg, W L C Wheaton and C B Wurster *Explorations in Urban Structure*, Philadelphia, PA: University of Pennsylvania.

Welford, R J (1995) *Environmental Strategy and Sustainable Development*, London: Routledge.

133

Part III
Cultures of Infrastructure and Public Space

CHAPTER EIGHT

From Café to Park Bench: Wi-Fi® and Technological Overflows in the City

Adrian Mackenzie

> *Wireless: technology that permits the active transfer of information involving emanation of energy between separated points without physical connection.*
>
> (Wolfowitz 2004)

Mobile digital communications connect movements of people to movements of data. Wi-Fi, a 'wireless technology' in US Assistant Secretary of Defence Wolfowitz's terms, connects people and data in a quite complicated cultural-technological mix. In certain contexts, it dispenses with the wires that connect computers and the Internet. But this mundane innovation in new media infra-structure seen in many domestic, commercial and organizational settings, can be interpreted as significant in different ways:

> [T]he real value of a satellite television broadcast, a WiFi connection to a laptop, or a mobile phone call from your car to your mother isn't the absence of dangling wires. Mobility, portability, ubiquity, and afford-ability are all enhanced when signals pass through the air rather than through strands of copper or optical fiber.
>
> (Werbach 2003)

This chapter argues that the values of 'mobility, portability, ubiquity and affordability' involve a mutual contextualization of movements and images of movement. Put more baldly, the plethora of figurations, practices, commodifi-cations, pricing-models, gadgets and modifications associated with Wi-Fi can be seen as an image of movement as well as an infrastructure that re-positions people in relation to movements of data. The analytical problem confronting evaluations of heavily imagined mobile technologies is how to hold together images of movement and movement itself.

Off the Lead in Piccadilly

In November 2003, Broadreach, a London-based Wi-Fi Internet Service Provider (ISP), opened a free Wi-Fi 'hotzone' centred on Piccadilly Circus. This company is one of dozens currently seeking to commercialize Wi-Fi networks by offering pay-by-the-minute wireless Internet connections in various public and semi-public places such as railway stations, departure lounges, hotel lobbies, train carriages, service stations, fastfood restaurants, cruise liners, trailer parks, public parks and cafés (e.g. Abreu 2003; Glasner 2003; Smith 2004). In most cases, these services consist of a wireless access point providing high band-width Internet coverage up to several hundred metres for customers carrying a Wi-Fi equipped device (laptop, PDA, VoIP – Voice over Internet Protocol – mobile phone, etc.). In contrast to the more well-established and visible Internet cafés (Wakeford 2003; Miller and Slater 2000), customers use their own equip-ment and sit or move around according to what they are doing. By dissolving the physical connection of wires, Wi-Fi allows a re-positioning of some now mundane everyday practices associated with new media. Wi-Fi networks them-selves are usually formed on an ad hoc basis, since people come and go, making and breaking connections to the network. Because of their limited coverage, these small networks are usually called 'hotspots'.

Since their inception in 2002, the cost and difficulty of logging on to these hotspots has meant they have not been heavily used. Dire predictions of another dot-com style collapse have abounded in IT and business pages of print and on-line media:

> Hopes that the roll-out of wireless broadband networks – so-called wi-fi hotspots – will result in a profits bonanza will be dashed, the tech-nology consultancy Forrester has warned. 'With all the hype today about the rollout of . . . public hotspots, it's as if the dot.com boom and bust never happened', said technology analyst Lars Godell.
>
> (Weber 2003)

Despite, or perhaps because of, these predictions, many efforts are being made to expand the coverage and extension of these Wi-Fi hotspots, to extend them beyond the precincts of the café to include the surrounding streetscape or adjacent open spaces. In Broadreach's case, a new degree of spatial exten-sion was achieved by linking adjacent hotspots located in different cafés, offices and bars together to form a *hotzone*. Their achievement had been to extend the spatial extension of the network: '[t]he zone stretches from the east end of Piccadilly, from Church Place onwards, through Piccadilly Circus and down Coventry Street as far as Wardour Street' according to the Broadreach press release (Smith 2003). This significant location in London connects a central node of the tourist geography of London to a street associated with new media, advertising and fashionably high-tech companies.

In late November 2003, at the height of the Wi-Fi boom, I was sitting on the steps of the monument in Piccadilly Circus with a laptop computer. In the morning, I had interviewed an artist using Wi-Fi networks to broadcast 'local television' in the park at Bedford Square. Embarrassed to be using a laptop on steps covered with tourists and lunching city workers, I spent only a few minutes trying to locate the Broadreach hotzone. Given that the whole area was well within the hotzone, and that access to the network was supposed to be free for that month, it seemed like a good opportunity to see what it would feel like to connect to the Internet in the midst of the London buses, the taxis, the barrage of signage and flows of people moving towards Leicester Square and Charing Cross Road.

There was no Wi-Fi signal. I could make no connection. Disappointed, I shut the laptop and walked down toward Leicester Square. But my failure to 'find the network' in one of the most networked zones in the UK is not the end of the story. I headed for lunch in a café closer to Charing Cross Road. A huge lunchtime crowd of people were flowing in both directions, going in and out of shops, cafés and cinemas. A few metres ahead, a small dog, a Yorkshire terrier, was trotting along, weaving between people, occasionally looking back over his/her shoulder. The dog was clearly not a stray or a street dog. It was well fed and groomed, but a little anxious. Sometimes it veered off, as if to take a turn up a side street, but then it moved back into the main stream of people. Sometimes it sped up only to slow down again. Not only couldn't I connect to the Wi-Fi hotzone, I was starting to feel responsible for a small lost dog. I was on the point of calling to the dog, when it looked back over its shoulder towards the other side of the pedestrian area. Strolling on that side, parallel to me, was a casually dressed man carrying a dog lead. He steered closer to me. In a friendly way he said something like 'she knows where she's going'. Feeling somewhat foolish yet again, I turned into a café, which I think must be part of the hotzone network, and tried again to connect to the network. When no signal appeared, I gave up and ordered lunch.

Folding Data into Places

What lesson can be drawn from this unsatisfying anecdote of Wi-Fi in the city? In respect to the dog, the connection was hard to see because of the absence of leads (or wires) and because of the flow of people that separated dog and human. The dog was not on the leash. This, the man might have said, is because a leash can get tangled in the movement of a crowd. Moving through a crowd with a dog on leash, even an 'obedient dog', is harder than moving independently. But how did the dog maintain a connection to the movement of the accompanying human? What I saw as 'being lost', as uncertainty or even anxiety in the crowd, was, for the dog, a relatively continuous, visual co-ordination of its path with a companion human. Is it taking the parallel too far to ask whether the contemporary problem of moving physically while keeping the network connection to everything we do is comparable to the watchful, anxious movement of a dog through a crowd of people?

The problem this chapter addresses is how to think about what is happening to networked communication after the dot-com crash, that is, in the ruins of projects, academic or otherwise, that treat digital communication as situated in a space apart from mundane, everyday, geographical and geopolitical space (Lovink 2003). The network society of the 1980s and 1990s moved ever-mounting piles of information between points on relatively geographically fixed networks. These movements triggered the effects of virtuality that have been the object of such energetic academic, political and economic contestation over the last decade (Miller and Slater 2000). The dot-com boom was animated by the image of placeless, seamless flows of information. The circulation of capital through financial markets speculating on eCommerce was mutually con-textualized with images of data on the move magically pervading the texture of everyday life. 'Rather than being seen as technologies to be adopted and shaped within the fine-grained practices of everyday urban life, new media were cast thus in this dominant discourse as a dazzling light, shining above everyday concerns' (Graham 2004). One response to that dazzle has been to see attention to 'everyday concerns' or what people actually do as an antidote. However, what makes the analytical recourse to everyday life difficult to accomplish is the panoply of figurations of new media as changing 'everyday concerns'. The notion of everydayness has become particularly problematic precisely because it has become the focus of images of new media change. In other words, the lesson of the dot-com/virtuality conflagration has been learned not just by academic criticism.

In the last three years, wireless networks for computer communications have burgeoned in settings ranging from 'remote places' such as Everest Base Camp and Nepalese yak farmers, rural villages in Cambodia to the 'first wireless nation' (Nuie). But it is in everyday locales such as railway stations, trains, airports, public parks, cafés, schools and houses that these networks have been most often set up, used and figured. Wi-Fi networks have found myriad uses – streaming audiovisual materials throughout the home, tracking children in Legoland, monitoring the growth of grapevines or wildfires through wireless sensor networks, or, more ambitiously, replacing corporate-owned telecommu-nications infrastructure with community-owned communications networks (Picopeering Agreement). Incontestably, the proliferation of Wi-Fi networks afforded an increase in the mobility of digital data. But equally incontestably, the proliferation of Wi-Fi has been the object of mobilization of a vast number of projects, reports, opinions and enterprises of many different kinds.

However, unlike much more general phenomena such as the WWW, Internet Relay Chat (IRC) or email, Wi-Fi concerns the folding of data into specific places. It is as if the diversity of Wi-Fi projects and the production of images of Wi-Fi networks anticipate critical calls for an urban new media studies based on fine-grained practices of everyday life. To describe the flows of data on Wi-Fi networks, then, would be to move across many different urban places, to cross many boundaries between public and private, between institutions and commerce. As the data flows across many boundaries, they encounter many differences relating to global-local, individual and collective identity, political participation, media, economy and technologies. A phantasmagoria of critical new media consciousness, Wi-Fi, as a constellation of attempts, projects, experiments, marketizing initiatives and regulatory policies, generates constantly varying forms. These forms negotiate differences, uncertainties, obstacles in the 'fine-grained practices of everyday life'. Individuals, groups and corporations develop embodied, institutional or commodified forms that feed further imaginings of mobility, ubiquity, portability and affordability. Images of data moving and data in movement follow each other, like the dog and its walker moving through the streets of London.

City in the Hertzian Landscape

The desire to place new media technologies above everyday life in the city remains strong. For instance, in describing changes associated with wireless communication, the architectural writer W.J.T. Mitchell recently proposed the notion of the Hertzian landscape: 'Every point on the surface of the earth is now part of the Hertzian landscape – the product of innumerable transmissions and of the reflections and obstructions of those transmissions' (Mitchell 2003: 55).

The idea of the Hertzian landscape follows from a discussion of 'Hertzian space' in (Dunne and Raby 2001) and refers to a landscape of data transmission. Some parts of the Earth lie within cities. Other points on the surface of the Earth lie in the middle of the sea. On the Hertzian landscape, these distinctions do not matter. Mobile phone networks, television and radio transmissions, wireless data networks weave together to make a 'landscape' that overlays and overflows topographical, geographical differences between points. But this landscape, which has become a site of major commercial, cultural, military and regulatory contestation is not easy to map or manage. It lacks the visibility and stability of other forms of space and property. It is more like sea than land. To the extent that it escapes full regulation, it is a smooth rather than striated space (Guattari and Deleuze 1988). A point on the geographical landscape spreads across different points on the Hertzian landscape, because different signal transmissions are present there.

In his account of the networked self in the contemporary city, Mitchell places great emphasis on certain kinds of networks exemplified by mobile phones and, less obviously, by the wireless computer networks popularly known as Wi-Fi. What is at stake in wireless networks, for Mitchell, is a transformation of public space induced by the movements of data between mobile points attached to different networks. The development of wireless networks, however, changes the stakes according to Mitchell. Now it is a question of 'reactivating' public spaces by bending network fixtures to suit different kinds of local movements:

> In general, as these transformations of public space illustrate, there is a strong relationship between prevailing network structure and the distribution of activities over public and private places. . . . And where networks go wireless, they mobilize activities that had been tied to fixed locations and open up ways of reactivating urban public space; the home entertainment center reemerges as the Walkman, the home telephone as the cellphone, and the computer as the laptop.
>
> (Mitchell 2003: 158; see also Sheller and Urry 2003)

While these changes are often presented as ubiquitous, it is immediately obvious that they are unevenly distributed. Wireless networking, the 'next big thing' after the Internet could easily re-inject another version of the oft-mentioned digital divide into the technoscape (Appadurai 1996: 34). The principal stake in the political economy of the Hertzian landscape can, therefore, be framed as a question: who today can send what? The flows of messages, images, texts, conversations, audiovisual sequences, control data and transactions that move through the Hertzian landscape are subject to many different obstacles, conflicts, competitions, and delays. The Hertzian landscape itself shifts as it undergoes rapid transformations. Countering the total vision of the Hertzian landscape posited by Mitchell, it might be possible to argue that Wi-Fi re-positions some existing communication practices in quite mundane ways. But in consonance with the affective tone of Mitchell's account, it is possible to see the actual projects and practices constantly overflowing or exceeding themselves in interestingly symptomatic ways.

Overflows

In Wi-Fi (or 802.11 Wireless Local Area Networks), data flows differently in the fabric of urban places. More precisely Internet Protocol (IP) data takes new paths, without physical connection, through built environments. In analysing this landscape of data flows (some urban studies take the form of showing that flows of people, things, images and data structure city space (Castells 2004: 85)), the notion of flow as shaping the city can be supplemented by a notion of *overflow*. The flows of data and the re-positioning of everyday practices associated with Wi-Fi are examples of movement that could, generalizing a notion proposed by Callon *et al.* (2002), be called 'overflow'. The notion of overflow incorporates the notion of flow and circulation as structuring the city, as folding data into places. But it entails certain forms of excessive flow, phenomena exceeding the scope and parameters represented and regulated within normative framings of networks as connecting people. By virtue of these overflows, identities, embodiments, feelings, affects, images, gestures and habits attach to flows. Put abstractly, the point would be that flows are never contained. They are relational entities. The ontological instability of flow consists in overflow.

The concept of overflow found in Callon *et al.* (2002: 287) suggests that technologies constantly create uncertainties in relation to identities, institutions, debates and practices. This is not due to any intrinsic property of a particular technology. Rather, it stems from the way technologies are enmeshed with markets and consumption. Callon writes:

> There's a strange meshing of techno-sciences and economic markets which produces what Marilyn Strathern calls the proliferation of new identities and which constantly creates new uncertainties about the constitution of the collective. So this constant creation and proliferation of the social (or what we propose to call emerging concerned groups) requires new procedures, new institutions, political institutions, new forms of debates and so on.
>
> (Callon *et al.* 2002: 287)

A related point had already been made by Appadurai when he wrote that movements and images of movement mutually contextualize each other (Appadurai 1996). In other words, flow as continuous movement and image as figure or form are integrally related (Gaonkar and Povinelli 2003). While Callon does not discuss images to any great extent, the main motivation to introduce the notion of overflow is to complicate the idea of flow as an effortless achievement or a quasi-automatic process, and to show that a flow can be intrinsically animated by an interplay between images and movements. While overflows may later disappear, every technology passes through a time during which an 'interactive stabilization' is needed to establish its boundaries.

For the purposes of illustrating how overflows give rise to flows, three different categories of Wi-Fi-related phenomena are useful: personal-infrastructural; technico-practical; and cultural-economic.

The hyphenated categories are not exhaustive but mark out zones of contestation associated with Wi-Fi. By their very nature, overflows are actually quite hard to separate from each other, but this categorization of Wi-Fi overflows provides a way to sort the sheer abundance of images, reports, accounts, advertisements and announcements in the years 2002–2005. Rather than simply making an equation between Wi-Fi and data mobility, the notion of overflows suggests that the plethora of concrete implementations, projects and practices are also an imaging-imagining of flow. It is this instability between image and movement that the concept of overflow helps to track.

Personal-infrastructural Overflows in Café and Park
Images of mobility figure data as moving effortlessly throughout the Hertzian landscape, and particularly throughout open urban places such as parks and

outdoor cafés (Toshiba Corporation 2003). To take two specific places, what is at stake in the mobility of digital data in cafés and parks? Both are places that have been significant in the life of cities. It would be possible to point to histories of cafés as spaces of contestation and dissent, and to mention their role in the formation of Western democracy (Anderson 1989; Habermas 1989). Parks and gardens too, have a significant socio-political history in the invention of publics and social movements that will not be recounted here. Leaving this aside for the moment, I am interested in seeing how movements of data through these urban places challenges both how we think about new media in cities, and what it might mean in terms of developing a relevant theorization of new media in everyday urban life.

Movement of data on the Hertzian landscape occurs only when certain boundaries are negotiated and a certain co-ordination between movements of bodies and movements of data are accomplished. It implies a changed relation between person and communication infrastructure. The key issue here is how movements of data can be reliably connected to particular individuals. With fixed infrastructures such as telephone networks, the systems of identification that attach points in the infrastructure to particular people are highly developed, place-related and regulated by contract or law. In principle, the technology of Wi-Fi poses no problems in terms of connections between people and infrastructure. It simply allows existing practices of computer usage to re-position themselves slightly. When a Wi-Fi network is installed in a café, a fast-food restaurant, a hotel lobby or departure lounge, it seems to change nothing. All it promises to do is allow people to carry out computer-mediated communication without having to plug computers or digital devices into a wall socket (telephone or Ethernet). Many media accounts of Wi-Fi hotspots in cafés present them in just such terms: Wi-Fi changes nothing, it is only an incremental step (Wainwright 2003). But this change has been remarkably difficult to accomplish in certain respects. In some contemporary parks and cafés, movements of people and data are connected, but only after changes in the relation between the person and the infrastructures of communication have occurred. Hotspots alter patterns of spatial, temporal and intersubjective relations, and impose new orderings of those relations. These orderings have to become more or less habitual before the connection between person and infrastructure becomes tenable. In short, the accomplishment of this re-positioning requires a large number of other changes to have occurred, and these are by no means straightforward and uncontested.

The management of this mundane re-positioning implies many other changes. My attempt to surf the World Wide Web from Piccadilly Circus relied on gestures and perceptions that are familiar habits, habitual within the space of office or home. The failure of this attempt to connect to data on the move could have been due to any number of reasons. It could have suffered from a mismatch between the broadcast strength of the Broadreach infrastructure and the antenna sensitivity of my laptop computer, or it could have been a failure on the part of the software to pick up the SSID (the Service Set Identifier)

Broadreach was using, or it could have been that I was sitting at the wrong angle on the steps. These possibilities leave aside the increased complications that exist when, as usual, the hotspot is not free of charge. The complications here include high cost, credit card verification difficulties, competing hotspot pricing schemes, and time-based usage schemes that constrain connectivity in particular ways (Macdonald 2004). Such failures are symptomatic. They highlight the diverse kinds of connections and co-ordination work needed to ensure that infrastructure is readily available and relatively invisible. The status of this work is unstable. On the one hand, if it is not done, nothing works. On the other hand, if it is too visible or too explicitly technical, then infrastructure itself becomes an obstacle. As it becomes visible, it no longer functions as an infrastructure.

Other aspects of this kind of personal-infrastructural overflow are easy to find in the context of Wi-Fi networks. Problems of shared infrastructure obsess ISPs (Internet Service Providers) who want to stop their bandwidth leaking into the hands of data-hungry masses. Because Wi-Fi networks overflow physical boundaries such as walls and windows, they can be accessed by unknown others. In 2002, a minor blaze of media interest was ignited by chalk-marks found on streets in London, San Francisco, New York and Melbourne (N/A 2004; Hammersley 2002). These marks indicated the availability of freely accessible Wi-Fi networks. While the practice of 'war-chalking' was itself relatively short-lived, war-driving and war-flying continue the practice of finding, using and sometimes publicizing the availability of Wi-Fi networks in urban spaces.[1] Many efforts to map available Wi-Fi networks, and to publicize their locations through websites exist. In relation to infrastructure, the established differences between self and other have become complicated. For instance, does an 'open' wireless access point mean that its owner is happy for anyone nearby to make use of it? War-chalking websites syndicate information about open wireless access points:

Thu Apr 22nd, 2004 at :13:27 AM EST
Plant yourself in The Lemon Tree just off Trafalgar Sq and access the interweb for free, sitting at the window. There is an open access point that hasn't been chalked yet. It has nothing to do with the pub but thought you'd may as well have a pint! Enjoy. n.b. This point was open on Friday night 16th April.

(Hero 2004)

The ethics and legalities of sharing the wireless network inadvertently left open by someone in the neighbouring apartment or office block are complicated (Cohen 2004). On the one hand, it could be seen as bandwidth theft, and certain broadband ISPs have sought to present it that way (Charny 2002). On the other hand, others propose it as ethical resistance to the security sensitivity that recently led the US Department of Homeland Security to announce that Wi-Fi networks constitute a critical threat to the national information infrastructure (Boutin 2002), and the US Department of Defence to declare that consumer-grade Wi-Fi equipment must no longer be used by soldiers (Jardin 2004; Wolfowitz 2004). The key point, however, is that the personal-infrastructure relation is being imagined in different ways. Different practices experiment with the altered spatial, corporeal, financial and organizational relations between person and infrastructure. The first kind of overflow thus concerns the way in which habits and familiar patterns of movement, whether of people or data, are constructed and contracted around infrastructural changes in communication technologies. This overflow, rather than changing the devices themselves, or resulting in a proliferation of modifications and repackagings, involves an interactive and simultaneous adjustment between person and systems in which various spatial boundaries and physical obstacles to movements of people and data are negotiated (compare Licoppe and Guillot, this volume).

The Techno-practical Overflow
The 802.11b technology was designed for office environments and local area networks confined to limited precincts within a range of a few hundred metres (IEEE 1999). But many projects and inventions have shifted the technical limits of 802.11 communication protocols devices in unexpected ways.

A first kind of technico-practical overflow involves modifications of the spatial extension of Wi-Fi networks. New kinds of antennae, often made from readily available materials such as Pringles cans, cooking equipment or old domestic satellite dishes have extended the range of connection to kilometres; new kinds of software have been developed that allow people to link Wi-Fi hotspots together so that data can leap-frog across wireless networks to and from wired or cabled infrastructures (Negroponte 2002; Batista 2003). In addition to the spatial overflow of the limited range intended for Wi-Fi, and the networking of Wi-Fi networks together in meshes of connections through various coordinating projects that include community networks and commercial wireless internet service providers (WISPS), the technical protocols for Wi-Fi are now being re-packaged in a gamut of devices and technical assemblages that include bikes (Gitman 2004), lamp-posts (Kewney 2004), cars (Best 2004), shopping-carts (Zuniga 2003), telephones (Datamonitor 2004) and myriad other devices. Many of these devices shift the border between infrastructure and the information that it carries. By making infrastructure more or less visible, information itself begins to move differently and to enact different kinds of collective mobility.

The practical affordances they may permit for their inventors or participants are less important than the ways in which they make flows of wireless data appear more labile, more potentially mobile, less constrained than the existing configurations of the technology presented in commercial productions, user manuals or commercially advertised wireless internet services. Scholarly analyses of new media technologies have justifiably criticized the hyperbole that imbues advertising and mass media images of technology (Morley 2003).

These images often promise freedom from constraint, and elimination of the impediments and obstructions posed by others. Strikingly, Wi-Fi has been accompanied by many imaging practices that diverge from this hyperbole, and take it in different directions. The *Consume* project based in Greenwich, London, seeks to syndicate community networks through a collaboratively constructed Web-based register of wireless access points (*Consume* 2003). By registering personal or group-run Wi-Fi access points, and showing them on an online map, the *Consume* project has tried, with varying degrees of success, to 'trip the loop' or construct an alternative infrastructure to that offered by commercial broadband internet service providers. *Consume*'s searchable community hotspot locator preceded the Web-based corporate hotspot locators run by Intel and other commercial wireless service providers. The utility of the *Consume* website and database is hard to gauge, but ideas of a mesh of publicly accessible wireless access points covering London quickly circulated through mainstream and on-line media. The *Consume* website, with its databases and maps presented an image of extended city-wide connectivity that exceeds and complicates the image of Wi-Fi as an office or building-limited network.

Collective-cultural-political Overflows
The final form of Wi-Fi-related overflow concerns the constitution of collective life in the city. The city and the political are closely linked for Western political traditions because special places in the city are significant to communicative praxis. These places, including cafés, coffee houses, squares, parks, malls, courts (in all senses), halls and houses, form the material fabric and infrastructure of the public sphere. Special places need to be rethought in the context of different patterns of movement and new forms of communication occurring in them. In exploring the implications of overflows, Callon asks:

> [I]f you consider overflows, you don't know who is concerned. Is it an individual? Is it a group of individuals? Is it a hybrid collective, mixing humans and non-humans? Nobody is able to answer this question. So it's a principle of uncertainty about what the collective is made of, or will be made of. The notion of citizenship is usually linked to the notion of an individual or society made of individual citizens who have to be integrated, and who have to be more active in order constantly to recreate and rebuild the social link.
>
> (Callon *et al.* 2002: 288)

In the National Mall, Washington DC, one of the most potently symbolic public parks for Western nation-states, the OpenPark Wi-Fi project intends to assist in the re-invention of democracy for the twenty-first century (OpenPark. net 2004). As in many other public parks and squares (Bryant Park (Bryant Park 2003), Central Park), this Wi-Fi project provides free Internet access for anyone visiting the park. However, here within the National Mall precinct,

Internet access is linked to democratic public criticism by assembled citizens rather than simply an enthusiasm for consumption of online information. In their press releases and website, the OpenPark project somewhat optimistically suggest that the democratic mechanisms of representation can now be rejuvenated through flows of Internet-enabled public criticism: '[f]ree hotspots for democracy – so that citizens can communicate with each other online when exercising their 1st Amendment rights' (OpenPark.net 2004).

While it is hard to take OpenPark's hope to rejuvenate democracy through communication seriously in the face of other obstacles to political change, it is possible to treat the introduction of Wi-Fi in the National Mall more generally as an intervention in the constitution of the collective. The overflow, in this case, runs between technology and the political but, as in earlier cases, it involves an interaction or contextualization between flows of data and images of movement. In all the projects I have examined, and all the instances of Wi-Fi I have looked at, a practical accomplishment of data mobility is closely coupled with an image or representation of freedom of movement. In fact, it would be possible to say that rather than being practical projects, everything I have discussed has been in some sense an image of movement. In these images, practical accomplishments such as a Wi-Fi network in a public park, a new kind of antenna that extends coverage, or a different way of connecting networks together through a web-enabled database, appear as figures in a scene that portrays some shift in identity, relations between people, or way of inhabiting a place. OpenPark is no different, except that in this case, a constitutional freedom of speech associated with people assembling in a public place is linked to the movement of data through that place.

A final example makes the point most directly. The ParkBenchTV project, installed by the artist Pete Gomes in Bedford Square, Bloomsbury, London, seeks to provide local TV coverage for people who habitually sit on the benches in the square (Gomes 2003). Existing television signals already cover this spot, but these signals are national broadcasts. ParkBenchTV broadcasts local content, preferably relating to the Bloomsbury neighbourhood, from a Wi-Fi antenna on the roof of a nearby building. It repositions TV audiences, shifting them away from the living-room sofa onto the public space of the square. In general, what I am calling a collective-cultural-political overflow in this context concerns how there is no possibility of conclusively defining the forms, the places or ways in which a Wi-Fi network can re-position existing communicative practices without taking account of the structuring of emotions, feelings, beliefs and perceptions concerning the act of communication. It is this that ParkbenchTV highlights most succinctly by turning Wi-Fi into the platform for a new version of an existing media form.

The question is whether changes in the forms of communication and the kinds of flow associated with communication can be understood in terms of existing institutions of political life within the city. Do projects such as OpenPark, and even perhaps ParkBench miss some of the uncertainties, the impossibility of knowing who today sends what?

Conclusions

Manuel Castells writes that: 'Moving physically while keeping the networking connection to everything we do is a new realm of the human adventure, on which we know little' (Castells 2004: 87). The problem of how to keep data moving in a way that is synchronized with the movements of a person is leading to the development of many different kinds of habit, anticipation and systematizations of mobility (Thrift 2004).

One analytical response to this situation is to introduce a new theoretical abstraction to explain communicative praxis. The idea of the Hertzian landspace is one such response. Similar ideas run through much futurological and policy work on telecommunications today (Werbach 2003). Such responses risk making the same 'mistake' made by earlier responses to the Internet. New patterns of information movement are treated as detached from, or of a different order to, existing forms and practices of everyday life. A more critical response draws on geography, sociology and cultural studies to argue for a new hybrid discipline, 'urban new media studies', which would locate movements of data in relation to everyday practices in the city. The nexus of 'urban' and 'new media' already signals a localization and specificity of analysis that the Hertzian landscape mostly lacks.

In some respects, the analysis of overflows developed here heads in that direction. However, the question is whether locating flows of information and data in the city goes far enough. The problem of analysing communicative flows in the city is complicated by the history of new media, a history replete with images and imaginings of movement, perception and action. On the one hand, wireless networks are imagined as the 'next big thing' after the Internet. The feeling that the mobile Internet is 'what comes next' runs strongly throughout corporate, governmental and art projects associated with Wi-Fi. In this respect Wi-Fi recycles and remediates many of the same claims, beliefs, images, values and emotions associated with earlier new media and digital culture – the promise of pure fluidity, absence of obstacles or constraints. On the other hand, the three kinds of overflow discussed above mix images of movements with practical negotiations of movement within everyday urban settings. In each case, the image of movement without obstacles encounters practical obstacles, to which different responses, forms and social-technical formations arise. This is the chief problem that confronts urban new media studies – how to analyse the mutual contextualization of images of movement and movement itself, particularly when movement itself becomes an image.

Like the dog off its lead in Piccadilly Circus, Wi-Fi enhancements of 'mobility, portability, ubiquity and affordability' trigger complex negotiations between what can be seen and what can be done, between imagining and acting. In the course of the three years 2002–2005, wireless networks became unremarkable parts of everyday domestic, commercial and organizational infrastructure. But they also led to a remarkable process of variation and modulation of data connectivities that neither hype nor operational implementation alone contain. Data flow, this chapter has argued, needs overflow, proliferation or excess of identities, figures and practices. Mobilities do not always simply move, but function as images that channel further action.

Note

1   The terms 'war-chalking', 'war-driving' and 'war-flying' extend a cracking technique from the 1980s known as 'war dialing' that used a program to dial a range of phone numbers and recorded those that might be entry points to computer or telecommunications systems (Raymond 1996: 470, 477).

Works Cited

Abreu, E M (2003) *PluggedIn: Web access in the clouds coffee, tea or WiFi?* Reuters News, Sat. 15 February, 8:53 am. Available from http://www.reuters.com/.

Anderson, B (1989) *Imagined Communities: reflections on the origin and spread of nationalism*, London: Verso.

Appadurai, A (1996) *Modernity at Large: cultural dimensions of globalization*, Minneapolis, MN: University of Minnesota Press.

Batista, E (2003) *Mesh Less Cost of Wireless*, 13 February 2003 (cited 14 June 2004). Available from http://www.wired.com/news/business/0,1367,57617,00.html.

Best, J (2004) *BMW and HP to make Wi-Fi Beemers for High-flying Execs*, 5 May 2004 (cited 20 May 2004). Available from http://networks.silicon.com/wifi/0,39024669,39120476,00.htm.

Boutin, P (2002) *Feds Label Wi-Fi a Terrorist Tool*, 6 December 2002 (cited 14 June 2004). Available from http://www.wired.com/news/wireless/0,1382,56742,00.html.

Bryant Park (2003) *The Bryant Park Wireless Network* 2003 (cited 7 May 2004). Available from http://www.bryantpark.org/amenities/wireless.php.

Callon, M, Barry, A and Slater, D (2002) 'Technology, politics and the market: an interview with Michel Callon', *Economy and Society*, 31 (2): 285–306.

Castells, M (2004) 'Space of flows, space of places: materials for a theory of urbanism in the information age', in S Graham (ed.) *Cybercities Reader*, London: Routledge.

Charny, B (2002) *Cable Companies Cracking Down on Wi-Fi*, 9 July 2002 (cited 25 October 2005). Available from http://news.com.com/2100-1033-942323.html.

Cohen, R (2004) 'The Ethicist: Wi-Fi fairness', *New York Times*, 8 February 2004.

Consume (2003) *Consume the Net* 2003 (cited 4 December 2003). Available from http://www.consume.net/index2.php.

Datamonitor (2004) *Skype: giving wireless PDAs a new voice*, 9 April 2004 (cited 12 April 2004). Available from http://www.theregister.co.uk/2004/04/09/skype_voip_pdas/.

Dunne, A and Raby, F (2001) *Design Noir: the secret life of electronic objects*, London: August/Birkhäuser.

Gaonkar, D P and Povinelli, E A (2003) 'Technologies of public forms: circulation, transfiguration, recognition', *Public Culture*, 15 (3): 385–397.

Gitman, Y (2004) *Magic Bike: wireless access bike* (cited 4 June 2004). Available from http://magicbike.net/about.html; now see http://yg.typepad.com/about.

Glasner, J (2003) *Trailer Parks Convert to Wi-Fi*, 27 May 2003 (cited 12 March 2004). Available from http://www.wired.com/news/business/0,1367,58784,00.html.

Gomes, P (2003) *Welcome to Park Bench TV!* 2003 (cited 25 October 2005). Was available from http://www.mutantfilm.com/parkbenchtv/.

Graham, S (2004) 'Beyond the "dazzling light": from dreams of transcendence to the "remediation" of urban life: a research manifesto', *New Media & Society*, 6 (1): 16–25.

Guattari, F and Deleuze, G (1988) *A Thousand Plateaus: capitalism and schizophrenia*, London: Athlone.

Habermas, J (1989) *The Structural Transformation of the Public Sphere: an inquiry into a category of bourgeois society. Studies in contemporary German social thought,* Cambridge, MA: MIT Press.

Hammersley, B (2002) 'Working the web: warchalking', *The Guardian,* Thursday 4 July 2002.

Hero, Anonymous (2004) *Plant Yourself in the Lemon Tree* 2004 (cited 25 May 2004). Was available from http://www.warchalking.org/section/pics.

IEEE (1999) IEEE Std 802.11b-1999 Part 11: Wireless LAN Medium Access Control (MAC) and Physical Layer (PHY) specifications: Higher-Speed Physical Layer Extension in the 2.4 GHz Band, New York: Institute of Electrical and Electronics Engineers, Inc.

Jardin, X (2004) *Wartime Wireless Worries Pentagon,* 26 May 2004 (cited 28 May 2004). Available from http://www.wired.com/news/politics/0,1283,63604,00.html.

Kewney, G (2004) *Wireless Lamp Posts Take Over World,* 15 January 2004 (cited 14 February 2004). Available from http://www.theregister.co.uk/content/69/34894.html.

Lovink, G (2003) 'Hi-Low: the bandwidth dilemma, or internet sagnation after dotcom mania', in *Dark Fiber. Tracking critical internet culture,* Cambridge, MA: MIT Press.

Macdonald, N (2004) *Wi-Fi in the Real World – pt. 1,* 11 February 2004 (cited 23 February 2004). Available from http://www.theregister.co.uk/content/69/35461.html.

Miller, D and Slater, D (2000) *The Internet: an ethnographic approach,* Oxford and New York: Berg.

Mitchell, W J (2003) *Me++. The Cyborg Self and the Networked City,* Cambridge, MA and London: MIT Press.

Morley, D (2003) 'What's home got to do with it? Contradictory dynamics in the domestication of technology and the dislocation of domesticity', *European Journal of Cultural Studies,* 6 (4): 435–458.

N/A (2004) *Warchalking: collaboratively creating a hobo-language for free wireless networking.,* 25/05/2004 – 17:26:40 (cited 12 May 2004). Was available from http://www.warchalking.org/section/pics.

Negroponte, N (2002) 'Being wireless', *Wired Magazine,* 10 (10).

OpenPark.net (2004) *The Open Park Project (Open Park): WiFi internet access on the National Mall 2004* (cited 12 June 2004). Available from http://www.openpark.net/.

Raymond, E. (1996) *The New Hacker's Dictionary,* Cambridge, MA: MIT Press.

Sheller, M and Urry, J (2003) 'Mobile transformations of public and private life', in *Theory, Culture and Society,* 20 (3):107–125.

Smith, T (2003) 'London gets UK's first Wi-Fi "hotzone"', *The Register* (cited 9 December 2005). Available from http://www.theregister.co.uk/2003/11/10/london_gets_uks_first_wifi/

——— (2004) 'Wi-Fi in the real world – pt. 2', *The Register* (cited 9 December 2005). Available from http://www.theregister.co.uk/2004/02/10/wifi_in_the_real_world

Thrift, N (2004) 'Remembering the technological unconscious by foregrounding knowledges of position', *Environment & Planning D: Society & Space,* 22 (1): 175–191.

Toshiba Corporation (2003) 'Enter the World of Freedom', Computing Promotional Brochure.

Wainwright, M (2003) 'The future is nearly in sight', *The Guardian,* 31 July 2003, 21–22.

Wakeford, N (2003) 'The embedding of local culture in global communication: independent internet cafés in London', *New Media & Society,* 5 (3): 379–399.

Weber, T (2003) *Wi-Fi will be 'Next dot.com Crash',* BBC News UK Edition, 2003/06/20 (cited 12 December 2003). Available from http://news.bbc.co.uk/1/hi/business/3006740.stm.

Werbach, K (2003) *Radio Revolution. The coming age of unlicensed wireless.* Available from http://werbach.com/docs/RadioRevolution.pdf.

Wolfowitz, P (2004) Directive number 8100.2 Use of Commercial Wireless Devices, Services, and Technologies in the Department of Defense (DoD) Global Information Grid (GIG), edited by D. o. Defense.

Zuniga, R M (2003) *The Public Broadcast Cart* (cited 12 March 2004). Available from http://www.ambriente.com/wifi/index.html.

CHAPTER NINE

ICTs and the Engineering of Encounters: A Case Study of the Development of a Mobile Game Based on the Geolocation of Terminals

Christian Licoppe and Romain Guillot

Introduction

For a long time, spatial mobility was just a means to an end. Even in the eyes of such a keen observer of cities as Simmel, industrial life led to a lengthening of distances, 'which makes of every useless wait or travel an irretrievable time loss' (Simmel 1989). Recently, social science studies of mobility have taken a different turn, arguing that mobility patterns should be understood with respect to an actor with motives, skills and instrumental resources pertaining to mobility, moving in an environment that 'affords' mobility in many ways. In this 'paradigm of mobility' (Sheller and Urry 2006b), places cannot be considered independently from the people that inhabit them, however fleetingly, and urban movement may be a creative experience by itself.

This 'mobility turn' in social sciences is particularly relevant to the current development of mobile technologies for handheld devices based on geolocation. Most of these technologies augment the urban environment of their users with geolocated informational resources that can be retrieved from the handheld device (supposedly equipped with the adequate software) when its user is close to them. Geolocated technologies are therefore designed to reshape the experience of mobility in urban settings. Some of these systems are also endowed with capacities for mapping the geolocated entities, such as virtual resources and persons (through the mediation of their handheld devices), and making such continuously evolving maps available to users. In that case, the screens of the mobile devices become a kind of public space in which connected users are, under certain conditions, mutually aware of their positions and movements. One can say that mediated mobility patterns become public, which is an instance of how 'physical and cyberspace come together' (Castells 2001), with consequences we will discuss later.

While there have been some accounts of the development of such technologies in university settings (Griswold *et al.* 2003) and gaming contexts (Benford *et al.* 2003), those were only experimental. This chapter presents a case study of the development of a game aiming to use geolocation technologies to build a location-aware community of players,[1] which actually went beyond the experimental, and was marketed by a Japanese telecommunication operator (telco). We conducted several waves of interviews between 2001 and 2003 in the startup firm in Paris, which was developing the game. On two occasions, one of us (R.G.), who had software development process expertise, engaged in participative observations for several weeks, by taking part in some parts of the design process.

Such designers are trying to develop a complex and innovative technology bearing on many types of user behaviour (mobility, play, communication) in a rapidly changing business landscape. What is important to them is to give an orientation to their design, that is to stabilize a configuration that weaves together the definition and classification of a given technology, the criteria and methods to assess what constitutes proper tests and how to interpret their results, and a cognitive frame mapping the relevance of the rules, routines, algorithms, functions, modules and affordances that may be incorporated in its design. We propose to extend the notion of paradigm that Thomas Kuhn proposed to describe the functioning of 'normal' science (Kuhn 1962) and to call such a cognitive, practical and normative configuration a design paradigm.

An important feature of the design paradigm is the identification of possible effects of the technology that may be amplified with proper design. Engineers usually rely on simple deterministic models where the 'force' of the technology has an impact on some forms of user behaviour. When no particular model of force imposes itself, engineers have to interpret their technology, working back from available observable data on its use, much in line with some social science research which has stated that technology is like a text, whose interpretation is emergent and depends on the particular organization of the situations in which it is used (Grint and Woolgar 1997). It therefore does not have force by itself, but it may be apprehended as such in situations organized in a way that makes technology accountable as something endowed with a particular pattern of force and efficiency.

In the course of the design trajectory, three models would be invoked to define the technology, its effects, and the corresponding design paradigm. The first is founded on mobility. It measures the effects of the geolocated game technology with respect to the movements players make for the sake of the game. Proper design aims to stimulate these game-induced displacements. The second is founded on the commitment of the player to the game experience, whose consequences are measured with respect to what the player will pay for and how much he will pay. The third and last is founded on communication, whose importance was in a sense rediscovered by the engineers, who, much like many of their predecessors who had underestimated the importance of inter-personal electronic communication in Web services, had not anticipated it.

The communication paradigm relies on the idea that interpersonal com-munication technologies are considered as enabling generic possibilities for meet-ing, making contact, mingling, interacting and exchanging.[2] Society appears as a gigantic reservoir of potential interaction, which information and communi-cation technologies (ICTs) will have the capacity to tap. The observable multi-plication of mediated interactions, in turn, lends substance to the notion that there is a particular virtue or force proper to any communication device, which will get people to mingle, chat, write to one another and exchange viewpoints by their means. This particular approach, which relates the definition and use of mediating devices with the actual exchanges they support, stands out from other social science descriptions of artefacts and action, which usually also deal with different kinds of technologies. In these other cases the object may incorporate memory data into the environment (Norman 1988), materialize an algorithm which guides its appropriation (Norman 2003), or trigger routines; it may be described as an agent that translates an injunction or a program for action (Latour 1993) or as a conversationalist itself, within an ethnomethodological frame (Suchman 1987). A proper design of communication technologies will involve 'communication affordances' (Hutchby 2000), whose effects will be indi-rectly measured in the observable stimulation of mediated intercourse that can be related to them.

In the communication paradigm, the effects of ICTs are measured with respect to the electronic traces that mediated exchanges have left in various databases. This technology-mediated effervescence in social exchange is taken to testify to a kind of energy internal to social structure, a potential energy for interactions that ICTs are supposedly able to capture, materialize, liberate or stimulate. The debate over the effects of ICTs, therefore, closely crosses the discussion of the force of social links and the vitality of solidarity. ICTs are assessed according to the manner in which electronic exchanges appear to support virtual communities (Rheingold 2000) and specific forms of network sociability (Wittel 2001), or to preserve social capital, understood in this case as the sum of the resources that an individual can mobilize through his inter-personal contacts (Lin 2001). Within such a communication paradigm, social intercourse is, by principle, laden with positive value, much like mobility was in the case of the mobility paradigm.

Our case study will retrace the design trajectory from its initial stages, where the innovators were assessing the geolocated multiplayer game within the mobility paradigm (the game was a bit like a massive multi-player online role-playing game [MMORPG], but it would be played on handheld devices rather than PCs, and people would have to move to play), through many trials, leading to a simplified game with a very loose reference to Internet games, called Nido. Nido was eventually developed within a hybrid paradigm, combining features from the mobility paradigm and the communication paradigm, and which we will call the interactional mobility paradigm.

For the designers realized the importance for players of being mutually aware of their position, and therefore of the public character of their mobility. The evidence for this could be found in the text messages that players sent to one another, in two different ways. First, players talk about mobility was oriented towards the fact that their position and movements were available to others. Mobility patterns were not individual, but were addressed to actual or fictitious known and unknown players and made accountable to those others. Mobility could not be separated from the web of electronic talk, handling the problems, embarrassments and innovative behaviour associated with its public character. Second, because onscreen encounters elicited many text messages it was evident that artefact-mediated mutual awareness of mobility patterns mattered greatly to players and to elicit many text messages. Improving design could be seen at the time as a way of engineering such encounters.

In the following sections we introduce the game and its context; then trace the development of a geolocated multiplayer game; the switch to Nido and the move towards an 'interactional mobility' paradigm; and the implications of 'on screen encounters' for the public character of mobility. Because they are not only specific to this kind of game, but are also crucial to the development of geolocated technologies in general, we finally discuss the questions public mobilities and onscreen encounters raise from an interactional point of view (with respect to interaction in public spaces and civil inattention); from a moral point of view (the issue of electronic stalking); and in regard to urban behaviour (a management of encounters based on technology-mediated mutual awareness of proximity, which may have more to do with environment-adapted flexible opportunism than long-term planning).

The Game and its Context

Jindeo is a game that takes part of its inspiration from massive multi-player online role playing games (MMORPG). In Jindeo, the players are supposed to set up teams that are to accumulate resources to construct ever more powerful artificial intelligences (AI), by means of quests and interaction with other players. The major originality of Jindeo[3] is to try to move MMORPGs from the connected computer screen universe to that of mobile phones. Not only because players may log on to Jindeo either from a computer connected to the Internet or via a mobile terminal (phone or PDA) but also because the gameplay is based on the new possibilities offered by mobile terminal location tracking. The position of the player in the game space map representation is closely related to the position occupied by his body in geographic space via the geolocation of his terminal in the mobile network. The common distinction in online games between a game space that is 'simulated' electronically onscreen and in which the player projects himself through avatars, and a 'real' space where the embodied player moves around, becomes blurred with Jindeo. The purpose of this game is, in fact, to construct a game universe that maps the space of ordinary embodied experience: 'The originality of the game stems from the use of mobile phones to immerse the player in the most perfect virtual universe: reality.'[4]

Jindeo then adopts certain characteristics of 'lifesize' roleplays – where players are gathered to interact in conventional places (town, hotel, forest, etc.).[5] Jindeo moves away from this, however, as regards the scale of the game universe. Here, it is not a question of people meeting in a closed place that they take over for the duration of a weekend, but of making the entire town a game universe. Players are expected to be engrossed enough by the evocative powers of the cityscape, as translated in the gameplay,[6] to meet in places where they would not have any reason to go and to move when and where they would not have moved without the game. In addition, the designers of the game claim that the game's major feature is that it can be played simultaneously by players using computers connected to the Internet network and by players using mobile terminals. This requirement has played a significant role in their choosing to base their *gameplay* on the progress of teams rather than individual players (the latter being the standard convention in most MMORPG). It was, in their opinion, the only way of ensuring effective forms of gaming co-operation between players using PCs and players using mobiles.[7]

To understand how Jindeo's designers imagined game situations, it is interesting to go back to the fictitious narrative with which their large game presentation document started. Here is a significant extract:

Friday, 5.27 pm Rue du Caire, Priss undoes her bike to return home ... Going home on her bike did her good since she avoided the harsh, closed-in atmosphere of the underground. She put her large Sony earphones on her ears and prepared to ride across Paris all the way to the 13th district. She plugged her player into her mobile phone. Better to answer straight away than be swamped with ten messages ... She slows down and stops to look at her mobile terminal on the corner of Rue des Ecoles. She glances at the contextual mapping module which indicates over twenty-five players from the same team roaming in her current cell area! She had set the alarm to twenty and it had been a long time since it had last gone off. She hesitates a moment and then attaches her bike because it's better to be unencumbered for proximity action.

She tinkers a bit with the screen to see which of her co-team members are hanging around. She had been so happy to get accepted in this team, because they were reactive rather than aggressive, keen to thwart the moves of much more belligerent teams ... The music kept pulsating in her ears. She decides to review the situation more thoroughly and to find a suitable, more private corner to avoid being caught stupidly now she has got into a sensitive area. She looks carefully around her checking for any suspicious movements and moves away from Boulevard Saint

Germain until she finds an open building. Twenty 'T+s' ('Turing positives') remaining now in her current cell, the others must have moved to another cell. Twelve 'U-Labs' and five cops ('IABasta police'). No 'Mtraders', though they own that particular area ... The only nearby exchange zone was dedicated to the trade of protection modules. Priss decides to load three defence modules and two spy mechanisms: this is the maximum she can pack in any case.

The Priss narrative[8] weaves together most of the elements that make up Jindeo into a tightly integrated game situation description, focused on complex game-mediated encounters and coordinations in a public urban space. In its core, we have a player whose displacements in the city are continually alternating, split between two roles and two participation formats, that of the ordinary embodied passer-by (Priss, who is returning home by her normal route, crossing quarters populated with city-dwellers to whom she is politely inattentive) and that of the equipped player (Priss who moves through 'cells' belonging to identified teams, populated by both city-dwellers and players that she has to

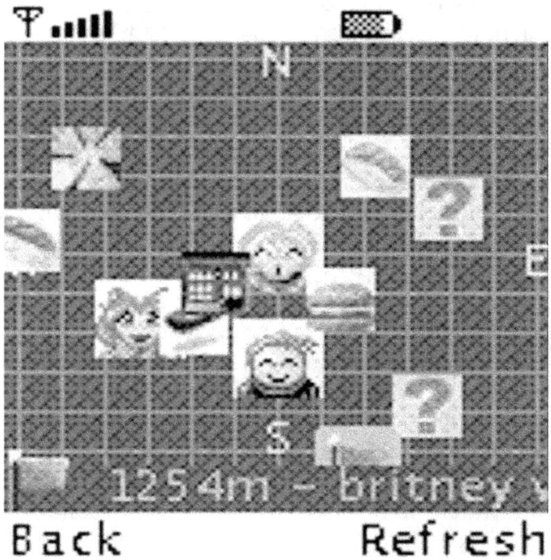

Figure 9.1 Nido Game Screen

The radar interface shows the local map of the game around the connected player (his icon appears at the centre of the screen). The square is about 1 km wide. Other connected players and geolocated virtual resources present within that distance appear on the screen. The nearest player in the buddy list is indicated at the bottom of the screen, with its current distance from the player, even if he is farther than 1 km. This functionality has been added by the designers to stimulate 'screen-based' encounters, as discussed below.

differentiate between, and which are revealed by her mobile screen). This participation format-switching is different from what may occur in many computer games, where actors connected for hours and hours on their computer may switch their attention and commitment continuously between professional activities and game activities by moving between active windows on their PC screen. Priss' urban space experience is different. It hybridizes action performed within the game frame and the urban context: there are no separate windows to switch between, they make up a single mediated reality she moves in and she lives in, even if her gaze may alternate between the street and the mobile phone screen.

Switching the attention between the mobile phone screen and the cityscape involves very distinct engagements into the situation. The player's perspective alternates between a subjective, embodied view of the town and an intangible, almost omniscient perspective. On the screen, he can see himself located on a stylized cartographic representation, which includes nearby members of other teams, material and virtual resources to be retrieved or exchanged in this sector, and fictitious obstacles to be circumvented. This permanent fluctuation of her attention between the mobile terminal and the urban scene allows Priss to differentiate between passers-by and players, between ordinary mobile phone users and players (themselves differentiated according to the teams they belong to), and adjust her actions accordingly. This particularly equipped situation is somehow reminiscent of that of an aircraft passenger who can alternately see the landscape passing beneath him and follow the progress of the plane on the maps displayed on the small television screens that surround him. But, in a plane, everything is done to ensure that the passenger with divided attention remains passive. On the other hand, in the game, the switches of perspective and attention of the actor engaged in urban activities are designed and exploited to empower the player to develop co-ordinated actions, meaningful from both perspectives.

For this purpose, the designers introduce numerous features into the design of the game, the use of which requires refined discrimination and co-ordination between the two perspectives. 'The wall' is one of them. It is an obstacle that only appears on the screens and has no referent in 'real life'. When the 'real' position of a player corresponds to the co-ordinates of the wall in the 'virtual' game map, and if the wall has been activated by a hostile nearby player, he gets 'frozen' on the map, until he has made a 'real life' displacement in the street, which is equivalent to a 'circumventing' of the wall on the screen. If a player thus has to walk publicly around an imperceptible and fictitious object only visible on the screen, he still has encountered it in some sense, and it has acquired a form of presence in the space of ordinary experience. From this example, we can see the extent of the hybridization of the embodied experience of the player and its meaning with respect to the game. Designers judiciously call situations of this kind 'game-related proximity engagements'.

In this particular interpretation of their game, the engagement of the player in the game and the pleasure he or she derives from it are reflected by the movements the player makes in relation to the game. The force of the game is determined by the scale, the difficulty or the number of such moves in actual space, that can retrospectively be read as 'induced' by the game. The farther those displacements, the more they seem to vindicate the game design in the eyes of its architects:

> We can envision gameplays in which a team is attempting to link Marseille and Paris by conquering squares in between; just think of the guy that gets on a train and stops at all stations to do so, or another which does the same thing from his car with his GPS, just for the sake of building up a chain belonging to his team between Paris and Marseille! Just think about it, it's amazing! It's absolutely crazy.
>
> (Michel, designer)

The design is then oriented towards stimulating many types of movements in the city, co-ordinated with the gameplay, either by inventing rules in the game scenario or by designing innovative interface affordances that would make such finely adjusted movements more compelling. Characteristically, when they were operating within this mobility paradigm, the designers thought of themselves as urban geographers and city planners, trying to incorporate the relevant abstract knowledge into the conception of their game: 'We're in the urban game, aren't we? We are almost urban planners. The problem is that we know nothing about it! If we knew more about pendular mobility, etc. it would have been really helpful' (Michel, designer).

The initial position of the designers was therefore to translate many features and values of internet-based multiplayer games into the real world of actual cities and mobile players. The specific force of the new game was initially conceived as its capacity to elicit game-mediated displacements because of the motivations (gameplay) and 'affordances' (localized virtual resources accessible locally with mobile terminals), which the designers made available to players. Conversely, such game-induced displacements could be used as evidence for the force of the game, thus providing specific criteria guiding successful design. At that time the game was therefore assessed within the 'paradigm of mobility' (Sheller and Urry 2006), in which displacements are less a means to a distinct end than a creative experience by itself, attuned to environmental resources. This is where the Jindeo concept was supposed to make a significant and even dramatic difference with respect to existing Internet MMORPG, where players almost never have to move physically to be able to play.

The Development of a Geolocated Multiplayer Game (2001–2003)

The Team

For the designers of Jindeo, the most evocative city is undoubtedly Tokyo, revisited by the aesthetics of the film *Blade Runner*, to which they often refer.[9] According to the 'bible', the basic game description document prepared by them, the idea for the game initially came to Michel R., already a MMORPG fan, when he was staying in Tokyo in early 2000, and still envisaging taking a job in the finance sector. He used his relations in Japanese banks to try the relevance of the Jindeo concept, first gathering opinions and then developing a business plan. A few months later, by mid-2001, these preliminary business discussions were useful for him to obtain funding to develop his mobile game concept from the Mobilfone mobile phone operator incubator, InnovaCell. The latter also provided the small Jindeo team with premises in the Paris Sentier district. This was the time when the New Economy craze was starting to fade, though the district was still being promoted by media coverage as being the Paris 'Silicon Alley'.

The incubator imposed very strict constraints on the composition of the team, refusing them to hire people on full employment contracts, and only authorizing them temporary hiring and various forms of outsourcing. This flexibility of employment systems caused the team to alternate between the passionate, focused commitment of young professionals in creating from scratch a permanent and fully blown firm, and the more flexible and revisable type of commitment usually associated with 'project-based' tasks. The shared passion for games, interweaving actual playing and designing practices, gave a distinct flavour to the motivation of all the participants, who apparently felt committed to a project in which their professional skills and their cultural engagements were aligned. Nevertheless, the exclusiveness of their attachment to the Jindeo project might be called into question at any time.

The nebulous organization that designs Jindeo revolves around ten people on the average. The designers who invented the original concept, and are now in charge of the evolutions in the scenario and game rules, are Michel who is the entrepreneurial soul of the whole thing, and Paul who got his share of fame when he developed in the 1980s a well-known 'space opera'-like roleplay game, before it was swamped by products derived from the worldwide fad for the Star Wars trilogy and its various derivative outputs in the game sector. On the technical front, the central characters are Pascal, the CTO, who heads the team of developers and Philippe, a later arrival, who was to take on growing responsibility at design level. There is a duality between the designers who act as the guardians of the founding spirit of the game, on the one hand, and the 'technicians' on the other, who constantly draw attention to the recalcitrance of the software systems to yield to the designers' impractical ideas (according to them). This duality shapes the organization of the office space. Besides a conference room reserved for meetings, everybody works in a single room, arranged in open space fashion, and divided into two groups of four desks set up in a petal shape. The creative designers are on one side and the developers on the other.

The difference between the two main modes of expression of the different actors – bursting, inspired, expansive creativity in the case of the designers, and stubborn, mute realism in that of the developers – is also obvious in the very different ways in which they set up dialogues within each group. Discussions between the designers look very much like oral jousting, each coming up with a new idea about the gameplay to resolve a previous objection. As the CTO mischievously puts it, 'fortunately, the storms and thunder of the creative designers have not contaminated the developers!' Quite the opposite, just next to them, the developers often absorb themselves into their screens. They even develop original methods of communicating with one another, annotating the bits of code they exchange in a co-ordinated software development workflow. Though in the same room, they rarely speak to one another. They exchange written comments through this silent and invisible electronic channel.

However the roles and postures of the various actors were to evolve considerably with the various trials encountered in the course of the design process.

Going for Japan and GPS-based Geolocalization (2001–2002)
The contract with the mobile incubator covered ten months' development, ending in July 2002. During this period, Mobilfone allowed an experimental access to its cell-based location-tracking server, to which requests could be performed through the WAP (Wireless Application Protocol) protocol. At the start of the contract, in the autumn of 2001, Michel R. was invited by InnovaCell to show his project in Japan, within the framework of a meeting of the WIVA (Wireless Venture Association). He took advantage of the opportunity to meet firms close to the Japanese operator, KDDI, which had announced it wanted to quickly launch services founded upon the location tracking of mobile terminals based on GPS (Global Positioning System). This location technique is much more accurate than those based on mobile phone network cell geometry.[10] In the context of a Japanese market well ahead as regards mobile phones with screens and Java graphic interfaces (much more user-friendly than the WAP interfaces available at the time in Europe), the Japanese also wanted services that would be accessible through such interfaces.

On this occasion, Michel met with the people from Tomen Telecom, a KDDI sub-contractor, which was looking for opportunities to supply KDDI with new, innovative services. They set up a meeting with KDDI management who confirmed the launch of GPS terminals in Japan and indicated their wish to develop game content based on the geolocation principle. KDDI showed a marked interest in Jindeo. After a few meetings, Tomen Telecom agreed to sign a partnership to manage the relationship with KDDI and take charge of the development of the Java interfaces for Jindeo, in exchange for an exclusive right to license the service in Japan, and a share in future revenue. The meeting with KDDI constituted a significant event in the development of Jindeo, imposing new constraints and leading to major evolutions of the initial concept.

First, KDDI required that the service provided by Jindeo should be accessible with national coverage, through graphic Java-based interfaces, as in all the mobile data services they commercialized. Leibsoft, who greatly needed a partnership with a major telecoms operator to boost its credibility, had to comply. This meant a radical evolution of the gameplay, for it had been designed with a focus on a particular situation, that of mobile players playing around in the centre of dense towns.

161

Second, KDDI offered the possibility to develop the game with more precise GPS-based geolocalization while the initial partnership with InnovaCell only gave the designers access to geolocation technology based on GSM cells, where the precision of this system was limited by the size of the GSM wireless network cells, i.e. a few hundred metres in cities. The enthusiasm was immediate. 'It's a dream come true, we can do Quake with that!' Indeed, within the mobility paradigm, in the mind of designers assessing the force of their game with the variety and subtlety with which players would achieve gameplay-mediated displacements, the precision offered by GPS-geolocalization (1–3 metres) seemed to open the door to a wide variety of playful proximity engagements, such as 'physically' taking shelter behind a 'real' building to avoid being directly in the 'virtual' line of sight of another player.

Third, the meeting was also the occasion of an experimental trial, for the designers had also set up a demo in Japan to convince KDDI. It served to dampen the enthusiasm elicited by the move to GPS terminals, by making the designers aware that localization was not all. A real setback lay elsewhere, in the time delays that went with each server request. In this trial, server responses required more than a minute each time. With such server response times, a given player may have walked or ridden around for at least one minute without his position having being updated on the screen. Even with a very precise location tracking system (such as GPS), the delays in server updates lead to irredeemable inconsistencies between the urban context and the onscreen context. These preclude many types of fine-grained co-ordinated movements, and put bounds to the force the game may attain within the mobility paradigm. That this was the case with Jindeo became obvious during the demos in Japan. Although the developers made a considerable effort in mid-2002 to optimize delay times and to rework their network architecture, the problem remains unsolved to this day.

The partnership with Tomen Telecom nevertheless gave the designers a welcome break. But the financial pressure kept on growing during the summer of 2002. The development contract with InnovaCell, with which relations had become strained, was supposed to come to an end. Leibsoft had sufficiently spread its investments to be able to maintain cash flow for several months, yet the need to find other forms of financing was more and more pressing. Under these accumulating clouds, Leibsoft approached a major video game publisher, Ubisoft, in September 2002.

The Ubisoft Trial and its Outcome (2002–2003)

In October, a presentation of the Jindeo project lasting several hours was arranged on Ubisoft's premises. This evaluation was carefully prepared by the designers, who attached great importance to this demonstration trial of their concept and architecture. Ubisoft is a major publisher, very well known in the sector, and had just won a contract with Sony for administration of the Everquest MMORPG game in Europe, acquiring still more prestige in French video game circles. The Jindeo designers were to meet with Ubisoft's marketing and game design specialists, including the new Everquest Europe Project Manager. According to the designers, it went as a complete disaster, following after previous negative comments from Wanadoo and Goa.

Though very cordial, the assessment of the Ubisoft game designers and marketing experts was focused on the market value of the game. How could the force with which a geolocalized game committed and engaged its players have market value? Why would players pay, how much would they pay, and to whom? The mobility paradigm was irrelevant in that respect. The game-induced displacements of players could not easily be distinguished from usual mobility patterns and accounted for. Even then, game editors and telecommunication service providers that were the major actors of the game value-chain could not conceive how they could make money out of physical mobility itself. Once mobility issues were put aside, Jindeo was just another kind of electronic multiplayer game, to be compared to MMORPG Web-based games, which were precisely the main field of expertise of Ubisoft. Two differences were then obvious. Rankings and reputation in the Jindeo game were attributed to teams and not individual players. Players directly accessed the full range of resources in the gaming universe, whether they were newcomers or individual players.

But in Web-based multiplayer games individualism and gradual entry into the game's virtual world are precisely the key to their economic value. Online players focus on increasing the skills, resources and score points of their own character or avatar. Such games are devised so that the player has to undergo a long individual learning process during the first levels of the game, a 'tutorial'. This long learning process is carefully organized around the progressive unveiling of an increasingly sophisticated universe. The length of such a tutorial also has the commercial function of developing the individual player's loyalty to the game due to the investments he or she has made to progress through its first levels. This is the point where the force of the game is applied, with the result of attaching the player to the game. It translates user behaviour into value, for it gets the player thus committed to the game ready to pay for further game-related features. Co-operation may however occur, but later, and with a lesser impact on marketing issues. Higher game levels usually foster some amount of co-operation, mostly in two different ways. In the game play itself, with a scenario and rules that link progression in the advanced levels to the trade of resources and skills. It also occurs through direct communication between players, in the galaxy of media dedicated to the game, whether official or amateur Internet sites. Forums and chats there are venues for intense interpersonal game sociability where people swap info (tips and solution tricks), download software to change or personalize the game ('add-on'), meet and organize with other players sharing the same values or the same ethos ('guilds').[11] But such games promote as a whole an individualistic and egoistic figure of a player, focused on the development of a character.

Jindeo was initially designed to stand out from MMORPGs for reasons that were ideological in nature but also related to the specificities of a geolocated mobile multiplayer game. Jindeo developers were, from the start, quite critical

of the individualistic nature of MMORPGs. Its developers definitely wanted it
to be something different:

> It is really different from other games inasmuch as there are no xp points
> [experience points]; it is much more community-oriented than current
> games, where everyone is more or less in it for himself. It is really team
> and community driven.

What is special about Jindeo is that the 'player does not progress in Jindeo,
only the team progresses'.[12] The player does not accumulate, 'every victory
or defeat is necessarily temporary and quickly no longer has any meaning
because it is diluted over time. Just as in real life', and 'the one who wins the
stake enjoys himself as much as the one who loses'.[13] The team progresses,
through the number of players that affiliate with it, through the number of cells
that have been claimed in its name and that the team controls (game cells represent a 'real space' segment of town) and the amount of software resources
accumulated by its members, and which are used to increase the power of
the collectively owned team AI, which it is the aim of each team to make as
powerful as possible. In addition, each team bears its own distinctive spirit,
ethos and set of values, which new players are expected to take on board and
subsequently enrich.

The particular stance of the Jindeo designers stems from a judgement of
value: they are seeking to counter what they perceive as a wrong individualistic streak in multi-player games.[14] But the choice of giving preference to team
progression rather than individual player progression in the gameplay has other
roots, besides criticizing the pervasive individualism of MMORPGs. It also
provides an elegant solution to the problem of attracting and interesting newbies
to the game. In online games, the newcomer only accesses the game universe
one step at a time according to his accumulated skills and experience points.
The purely virtual, immersive game universe can be split up into several distinct
sub-universes only accessible once the player has reached a certain stage, and
performed all the relevant quests at lower levels. Here the game universe is
largely 'real', for it is the city itself. Therefore, it is unique and difficult to split
up into distinct sub-spaces with differential access. Players will be able to move
in all of it, because they are part of it, and it makes up their whole environment.
You might restrict access to virtual resources, or the possibility of performing
certain actions or encountering certain players onscreen. That would, however,
reduce drastically the interest of the game itself, which lies in the multiplication
of possible co-ordination between onscreen actions and 'real space' actions.
So, to maximize the attractiveness, all players, whatever their actual personal
skills or the technical resources of the terminal they use,[15] should be allowed
as many encounters and actions as possible. If progress only occurs at team
level, the issue is nicely settled. Any player who gets close to a useful resource
may capture it for the team and help the team to increase its power. Participation
is made to depend more on *location* than on skill and experience.

The marketing people at Ubisoft were to criticize most severely this emphasis on team progression. They did not believe that such an incentive was enough to motivate players and make them loyal to the Jindeo universe. Based on their own experience of online games, they considered that if newbies were to have access to full player capacities in a single game universe, without going through a long and tedious individual tutorial, players would not become lastingly attached to the game. They would enter it, exit it, get back to it on a simple whim, unlike MMORPG universes where the tens of hours spent to learn the ropes, and get through the first levels were quite instrumental in building player loyalty.

What was at stake in the meeting with Ubisoft was not only the assessment of its economic potential but the very definition and categorization of Jindeo, and more generally of similar games based on the geolocation principle, where the game world is not virtual, but some kind of augmented real space. Because game-induced displacements were deemed to have no economic potential for a game editor (thus rejecting the mobility paradigm as irrelevant), Jindeo could only be compared to existing Web games on the basis of criteria built within the Web game industry and experience. Ubisoft criticism therefore amounted to saying that Jindeo was a new type of game that could not be assimilated to MMORPGs, at least as long as it worked that way. Moreover, it could not interest a Web game editor, because the very differences that made Jindeo special were running against the design features that made MMORPGs profitable.

That encounter with Ubisoft dealt Jindeo designers a severe blow at a time when they were too pressed to try to answer Ubisoft's criticisms and invent new rules and new scenarios that might do the job. Because they were, at that point, quite short of cash, they absolutely had to get something out fast, using only the working bits and pieces, software and architecture that already worked. This switch was both practical and conceptual, for it meant renouncing the initial project. This was stressed by the departure of Paul, the designer who had reached fame with a multiplayer role playing game in the 1980s and whose values could not be incorporated in the new design. The transition from a complex role playing game towards a simple collection game was not easy on the remaining designers either. The move to Nido seemed both radical and final. The designers did not even believe they might be able to get Nido to look more like Jindeo by enriching it with yet undeveloped modules and functions. There seemed to be an unbridgeable gap between an individualistic accumulation game such as Nido and an altruistic team-based game such as Jindeo. And part of the excitement seemed to have faded. In the first tests, its designers (immersed as they were in the massive multiplayer online game culture) found Nido rather boring. Designing Nido actually meant redefining the use value of the game and the exchange value that could be attached to it. We will try to show how it entailed a new design paradigm that paid attention to the interactional and social dimensions of mobility.

The Switch to Nido and the Move towards an 'Interactional Mobility' Paradigm

The Pivotal Role of Text Messages in the Redefinition of Design Paradigms

After the meeting with Ubisoft, the design space contracted to the existing and working modules, which were the Radar module, the collection module and the communication module. The Radar module tracked and mapped the location of the players and the resources present in a given cell, while the collection module enabled the player to collect available resources in the cell. The communication module allowed text messaging between players. Based on these three modules only, the simplest and possibly the only thing to do was to design a collection game, slightly inspired from successful image card collection games for children (*'Panini'*). The principle of this new game, Nido, was therefore to complete collections of virtual objects picked up with the mobile phones by clicking on the icon of such objects when they appeared to be close enough to the player in the onscreen electronic map. Some were collections of everyday objects, such as precious stones or fruit, and others did not have any real-world equivalents, such as moments of time. To collect those, one had to pick them at the right location and at the right time (you could only pick morning in the morning, etc.). The Nido onscreen map also featured the other players present in the cell. Finally, players could communicate with one another when their icons were apparent and active onscreen, via the text messaging module.

The 'bricolage' to which the designers turned was to become the Nido game, developed for KDDI and commercialized in late spring, 2003. Because of the simplicity of its gameplay, and the availability of several modules, the development of Nido got only one month behind schedule. The game was launched in Japan in early April, with limited visibility and almost no advertising. It was a moderate but reasonable success: 700 visitors to the site the first week, 50 subscribers and then stabilization at an average of around 200 players.

The designers were rather surprised to observe the level of involvement of some Japanese players in a game they thought simplistic and spiritless. Because the game was now publicly available, user behaviour became the main criterion to assess the effects of the introduction of particular design features. But user behaviour was mostly observed in one particular format. The innovators did not have enough money to conduct interviews with users. They could only rely on the electronic traces that user behaviour left in their databases. These traces could be either details of object collection for each user (number of objects, place and time of collection, etc.) or the text messages that users sent along through the game server as they moved through Tokyo and played Nido. Any significant modification of the game could be evaluated with respect to the number of messages it stimulated and the way it would be commented upon in the messages themselves. Since they based their representations of user behaviour on such traces, the designers could mostly link intensity of use to the proliferation of electronic traces, those left either by data collection or text messaging. Their reasoning, shaped by the particular constraints of the experimental setting, was mostly structured by the interpretation that the more numerous those traces, the more intense player involvement, and the stronger the force of the game, and the more relevant its design. This line of reasoning was quite compelling, not only because it was attuned to the only kind of available empirical evidence, but also because it offered a long-awaited solution to the economic questions raised in the Ubisoft trial: who paid what to whom? Text messages at least were paid for, to the telecommunication operator, for whom any way of increasing the mobile data traffic was an exciting perspective that would open doors, since it was part of their general strategy at the time. Designing the game in a way that players would play more and therefore send more text messages between them became very rational.

The first observations of user behaviour yielded two extreme behaviours: players interested mostly in accumulating virtual objects, and players oriented towards game-induced displacements and encounters. Some Japanese players thus developed a taste for pure accumulation, to the point of completing the same collection ten or 15 times. Some of these seemed to enjoy the game so much that they found ways of 'cheating', even though the game play was so rudimentary.[16] Though Nido appeared so unsophisticated and unexciting a game, some Japanese users were thrilled enough to invest a considerable amount of effort into the accumulation of virtual Nido objects. This particular type of involvement opened up an 'accumulation paradigm' for the design process. The design was oriented towards two directions. On the one hand the direct stimulation of *individual* accumulation by making objects and collections more attractive (a complete collection of a particular set of virtual objects might be rewarded with a free downloadable ringtone, or might compose in jigsaw puzzle fashion one of those starlet pictures the Japanese so much delight in,[17] to be used as a mobile screen saver). On the other hand, it meant getting these fetishist compulsive players to behave collectively by making it more interesting for them to trade. For instance, the designers introduced collections where objects were only regionally available, which forced distant players to trade. And trade meant players had to look actively for information from other players, which involved a lot of text messages.

In the second case, other players used the Nido collection game effectively as a pretext for wandering around Tokyo. This provided another path for a type of design oriented towards mobility, though it did not mean any longer trying to enable subtle co-ordinated moves between screen and city. The designers could just sufficiently spread the objects to be collected to allow for many pleasant 'hunting-gathering' trips throughout the Tokyo area. But their main discovery was that game-induced displacements featured heavily in the text messages. Players would comment directly on their displacements (all kinds of displacements, whether or not associated to the game) to other players, or orient messages towards the fact that some players might be aware of their positions. Text messages revealed the interactional importance of a feature of the game, which the designers had underplayed, that of making mobility patterns public, that is accessible to some other players in some circumstances. What would now matter were not individual mobility patterns, but mobility patterns that were addressed to others, and for which actual displacements could not be separated from the web of words in which they were embedded. We propose to call this 'interactional mobility'. It takes many forms, such as players guiding one another by text messages, chatting and joking about where they go. One of these forms, onscreen serendipitous encounters, would prove particularly important in the reorientation of the design process towards the interactional dimension of public mobilities, while being quite meaningful from a sociological perspective.

On-screen Encounters and the Public Character of Mobility
One of the key features of the Jindeo concept, which made such mobile geolocalized games so different from online games, was the possibility that players would meet face to face: 'I feel like saying, it's not like a game any more! To my knowledge, there is no (video) game that brings players directly into contact' (Antonio, developer). But such 'meetings' may occur in two distinct contexts, in real space (physical co-presence in the sense of being able to get mutual sight of one another) and on mobile screens (with the icons of both players being represented in the same cell map). What was still unclear in the initial stages was the importance of a particular configuration, which we will call on-screen encounters and which would stand out with the first Japanese users of Nido. On-screen encounters describe a configuration where two Nido players get to 'see' one another on the screens of their mobiles (without necessarily being co-present in the sense of a direct visual contact) and ratify this mutual perception on-screen as a proper encounter by commenting on it by text messaging.

The occurrence of such on-screen encounters contributes to the interpretation of the screen as a space of mutual awareness, where positions and movements are available to connected players whose identities cannot be determined in advance, which are present in the vicinity, and of whom the connected player will become reciprocally aware. In that sense, the screen is also a public space where the publicity of mobility is tried with every connection. In such an electronic public space, entities appear rather than reside on a permanent basis. Hannah Arendt described such spaces where 'persons do not simply exist as other living or inanimate things, but are explicitly appearing' (Arendt 1983). With the recurrent possibility of on-screen encounters, actors reflexively construe their mobility as accountable to other players. Within that frame, mobility is not individual but social in nature, and as the designers were to discover in the case of on-screen encounters, it requires a lot of interactional work.

An abundant text messaging activity was observed that was associated in one way or another (initiated by, commenting on, etc.) to mediated encounters. This allowed the designers to infer a relationship between the way their interfaces made players mutually aware of their positions and displacements and possible specific forms of mediated encounters, and the text messaging activity such encounters elicited. This would define a specific orientation for design, that is, providing interface 'affordances' promoting mediated encounters and shaping the way patterns of mobility were made public, with the beneficial consequence of stimulating the text messaging that went with such encounters. This is the essence of the 'interactional mobility' design paradigm.

The Nido players could be broken in two distinct groups, 30 or so KDDI employees that were given the system to interest KDDI in it, and about 200 'normal' subscribers to the game. The particularity of the KDDI players was that they all worked in the same building. Therefore, the concentration of connected players in these cells was rather high, and these users had many opportunities to 'meet' several other players on their mobile screen, even if they worked on different floors. They seemed to take so much pleasure in such mediated encounters that the number of relevant text messages exploded, highly exceeding the expectations of the designers. The emotion and pleasure they felt in these mediated encounters accounted to a large extent for the good reputation enjoyed by Nido in the Japanese firm. Such behaviour confirmed the initial idea of the designers that the attractiveness of such games would depend very much on the local density of connected players. However, the way interest in mediated encounters and subsequent exchanges superseded interest in the game play itself came as a surprise.

The 200 'normal' subscribers were in a very different situation. They were randomly scattered in the whole Tokyo-Yokohama conurbation and beyond. Therefore, they had very few opportunities for mobile phone mediated encounters, and they also communicated by text message much less than the KDDI group. This observation went a long way to confirm that there was a direct relationship between the number of game-related text messages and the number of game-mediated encounters. Mediated encounters could therefore be considered as a powerful lever to stimulate text-messaging practices.

The fact that users were scattered over a large area did not preclude two mediated encounters to take place, at least (others might have occurred without being mentioned in text messages). The first of these took place in the underground. In her train, a woman who used to play Nido but was not logged on at the time saw a man engrossed in his mobile phone, playing at Nido (or so she believed). She did not address him in any way, but got out of the train at her usual station. As soon as she was on the platform, she logged on to the game with her mobile phone. She checked that she could indeed see the icon of this active player on her screen. Only then did she send him a text message, thus ratifying the co-present encounter by this lateral and indirect communication. In the second mediated encounter, two connected players came across one another at night, in an otherwise deserted business district, while both were connected to the game. Their icons were mutually visible on their screens as they walked past each other. Since they were almost alone in the street, neither could ignore that the other one was that connected player that had appeared on their mobile phone screen map. But they just walked past without acknowledging the fact, either openly or electronically. Later, one of them recounted this particular encounter to another player by text messaging.

These two mediated encounters are very peculiar instances of those encounters in public space that Goffman has studied in detail (Goffman 1963). From the perspective of a bystander in the street, it might have looked like two unacquainted persons walking past, and granting each other some form of civil inattention. But, unseen from this hypothetical bystander, those two persons were meeting both as urban passers-by (to which the regime of civil inattention is relevant) and as Nido players. Their attention was prone to oscillate between the subjective and embodied viewpoint of the passer-by and the disincarnate, exterior perspective of the representation of the gaming plane on the mobile screen. Within that latter representation of the encounter, the mobile screens allowed them to recognize one another cognitively.[18] In these two particular instances, they chose not to make that simultaneous cognitive recognition mutual by failing to ratify it through direct social recognition.[19] In the first one however, social recognition occurred by the means of text messaging (a communication back channel), which turned the spurious co-presence into an encounter, in a way that could not be detected by a bystander. In the second case, social recognition was not proposed, but the SMS mention of that particular crossing to another party, meant that one of the participants at least interpreted that fleeting social gathering as a situation involving two 'acquainted' persons. There is a sense in which one could say that the fact these Japanese users were equipped with Nido, and logged on to it, turned the chance meeting of two strangers in public spaces (which would normally be treated in the regime of civil inattention) into fully blown encounters between 'acquainted' persons, with social recognition being accounted for and visible only in text messages. At least that was the interpretation of the designers, and we see that it makes even stronger the alleged correlation between the proliferation of text messages and the power of the game to subvert and transfigure urban social gatherings.

All these observations of user behaviour were discussed at length by the design team. They led towards the reorientation of the design trajectory in the summer of 2003. The innovation effort now aligned with the optimization of a technology that could be seen as a game, but also as a geolocated mobile instant messaging service, in which mobility patterns were made public, and whose map interfaces allowed for spontaneous on-screen encounters. The criteria for successful design became the ability of the interface functionalities to stimulate and regulate such 'encounters' mediated by mobile location tracking to enrich the corresponding experience and elicit text messaging activity. This meant for example making easier the sending of text messages, or creating 'buddy lists' that would allow friends and personal relationships to recognize one another on-screen:

> There are not enough places in the application where I'm tempted to send a message, or where I have the possibility to do it, and we missed a bunch of functions in our initial versions of the design such as displaying the nearest players connected. Now we are creating a list of

favourites, a basic Messenger thing . . . being able to see easily if my favourites are connected and close or not.

(Michel, designer)

A key feature to promote encounters was to ensure at least another player would be visible on the screen map. The innovators therefore added an additional feature, in which the icon of the nearest connected player in the buddy list appears together with his distance to the player, even if they are not in the same cell. This was thought, probably rightly so, to be a way to afford mediated encounters in contexts were there were too few players for their having a great chance of finding two connected ones in the same cell at a given time.

The design trajectory had therefore largely reoriented towards the conception and implementation of interface features that would promote and 'afford' mediated meetings *in general*. The engineering of encounters is a key step in the move towards the interactional mobility design paradigm. Mobility patterns are not apprehended solely as a set of game-induced displacements, but as a public activity, of whom many known and unknown connected players might be aware through the game interfaces. It is an unavoidably social activity, in the sense that situations involving mutual awareness and recognition of others' positions did elicit a lot of electronic talk.

Conclusion

On the analytical side, our case study has revealed two distinct representations that bore influence on design procedures and evaluation of technology. In the initial stages, the game was thought to provide motives for the player to move (through the gameplay), an urban environment augmented with geolocated resources (virtual objects and players' avatars), and software-laden mobile devices to act on them when in their vicinity. The force of the game was thought to materialize in game-induced displacements involving numerous coordinations between what the player could see around him and on the screen of his mobile device. Design was oriented towards providing geolocated features and software resources to enhance the capacity of the game to such displacements. This design paradigm was attuned to the 'mobility paradigm'.

In the later stages of the design trajectory, mobility was understood very differently. The designers had realized that a key feature of their game was to construct a space were connected players could often be aware of their mutual location. The game therefore provided a kind of public space, in which players reflexively oriented their behaviour towards the possibility that their displacements might be perceived and they might be called to account for them. Mobility could not be considered any more as an individual activity. The movements of connected players were always addressed directly or indirectly to the possibility of actual or imagined known and unknown players being aware of them. Mobility was therefore put into words, was discussed and commented in numerous text messages between players. A particularly significant feature of mobility patterns in such a mediated public space is the possibility of serendipitous on-screen encounters, in which a player discovered other players on his screen upon connection. Players proved to be extremely sensitive to such encounters, which were marked with particularly intense text messaging activity. Within this paradigm of socialized or interactional mobility, the text-messaging and communicative activity of players became the measure of the impact of the game. Design strategies turned towards ways to stimulate exchange between players, among which was the engineering of on-screen encounters. They tried to entrench interactional mobility in the very design and affordances of the geolocated game.

While these two design paradigms can be analytically distinguished, the path that led from one to another made for a complex and eventful history, marked by many trials. The technology and design were in a state of flux between the cash situation, technological and development constraints and market assessments. As most innovators, the game designers were trying to build a network of heterogeneous links that would stabilize their game, but several attempts at enrolling allies had dramatic consequences. One of the main steps in the move from one version of the game and design rationale to another was the attempt to enlist a big Internet game editor as a partner. Controversial debates in the resulting confrontation with marketing experts showed that mobility in itself had no economic value for the main actors in the game value chain, that is, game editors and telcos. Communication between players could be turned into profit, at least by telcos such as KDDI that were to market the simplified version of the game. Hence, the importance of considering how game-induced mobility was an addressed and socialized activity, and could be made more so.

Within this new paradigm for mobility, on-screen encounters based on the public character of mobility patterns became one of the cornerstones of the game development. We think such features raise new and important questions with respect both to the future of geolocated applications and social science, and that their relevance goes far beyond the particular case study discussed here. What kind of interactions are on-screen encounters? What are the moral consequences of the fact that position and movements may be available to a group of players? What kind of experience of urban environments is provided by systems that multiply the possibilities of on-screen encounters?

With respect to the interactional order some key notions, such as civil inattention, are at stake. Describing the behaviour of strangers crossing in large cities, the concept of civil inattention proposed by Goffman was founded on a principle of equivalence, in which every passer-by was the absolute equivalent of any other[20] (Quéré and Brezger 1992). That very equivalence is called into question with location-based services such as Nido, for even if the Nido-connected passers-by are not acquainted by name, some screen-based cognitive recognition relying on mutual placing of the other, both as a Nido player and as the connected player they both can see in the street and on-screen, is always a possibility. Is there still any such thing as socially insensitive civil inattention in a world where each actor is embedded in a personal environment saturated with Nido-like devices? The public space and the social order in such a world might be very different from the ones we inherited from modernity.

With respect to moral issues, a new set of questions appeared as the game evolved, dealing with the responsibility of the designers with respect to the morality of the encounters their technology mediated. This problem became more salient as the design rationale became focused on building an architecture facilitating generic encounters and socialization between any equipped user. Deviant encounters became possible, and likely. Indeed, one at least had already taken place. An adult Nido player had got into contact with a 12-year-old girl playing on her mobile phone. She revealed her age in the subsequent messages and he kept on flirting with her on a remote basis, at the same time following her movements in Tokyo from his PC.[21] This is quite unacceptable in Japan, where adolescent prostitution through the Internet is a matter of heated public debate. The designers realized that they had to take responsibilities for some of the risks involved in Nido-mediated mutual location-aware interactions, and particularly ill-intentioned forms of 'tracking' or tailing. Characteristically, the designers were looking towards the conception of interfaces adapted to do the job. For example, they were and still are exploring the possibility of providing users facilities for making up different lists, each involving different rights to mutual access and visibility, such as a list of on-screen mutually visible friends ('buddy list') and a black list that forbade such mutual visibility. This was only a start.

With respect to the urban experience, on-screen encounters extend possibilities of unplanned interaction to persons who do not meet face to face, but just happen to be close by and connected. In large cities where actual face-to-face encounters take a lot of planning beforehand, this could change the 'structures of anticipation' of actors (Thrift 2004). They could delegate to the mobile devices the burden of generating opportunities for encounters to happen, provided on-screen mediated encounters would happen often enough for them to be sure they would be able to arrange something at a convenient time. This situation is much akin to the experience of instant messaging, in which users rely on the state of an icon on their computer to determine whether their correspondents are somehow 'present' enough to make it worth it to test their availability for further interaction, thus turning a change of state in the icon into a full blown encounter. With respect to the organization of encounters in heavily mediated urban public spaces, a kind of flexible opportunism might therefore become more relevant than long-term planning.

Notes

1 For a classification of location-aware community systems from a user-centred perspective, see (Jones *et al.* 2004).

173

2 This is a recurrent feature in analyses of computer-mediated communication dealing with electronic mail, forums, chats, etc.

3 All names here are fictitious.

4 Jindeo Bible V0.3, p. 10.

5 'These multiplayer online games are, to some extent, inheriting several features of role plays on paper. Jindeo is in a way close to 'lifesize' role plays, I think' (Gilles, developer).

6 Jindeo Bible V0.3, p. 2.

7 One of the key references of the designers here was the first *Matrix* movie: they visualized players in front of their PCs being able to have a near-panoptical view of game situations (due to the power of the software and graphic interfaces on PCs) which would guide mobile players whose terminals have much less functionalities.

8 This is a reference to the film *Blade Runner*, where Priss is the name of a pretty, dangerous replicant.

9 The authors of *Blade Runner* were themselves avowedly basing their futuristic urban décor on today's Shinjuku district in Tokyo.

10 GPS-assisted geolocation does not actually use the GSM network as such, but satellite-based localization. Implementation of this technique requires the integration into each mobile phone of a GPS chip in addition to the classic SIM card. Because of the wavelengths used, a user cannot be located within a building or in a very dense urban area. This technology was available in Japan, and KDDI commercialized it. However, each localization request with this system cost five yen and took around ten seconds to be completed.

It is very different to the various geolocation techniques based on the GSM cell grid. Cell identification (or Cell ID), for example, is the simplest of all location tracking techniques. When the user is in a zone covered by the network, a single cell enables the connection and carries the call. It is easy to identify this cell but the accuracy of this type of location tracking depends totally on the size of the cells (around a hundred metres or so in a city, up to several kilometres in a rural area).

11 See the example of the game Diablo2 in (Largier 2002).

12 Jindeo Bible V0.3, p. 7.

13 Jindeo Bible V0.3, p. 6.

14   In all games, there is a real temptation to boost your character, often to the detriment of the team, and there is always someone who does this, though here, everything is designed to counter that tendency. A player must play for his team. He may leave his team but all he can do then is join another team. He gets more experienced as a player, more familiar with the rules, gets to know other players, but it is still his team that makes progress . . . The reflex consisting in trying to build your own character and observable in almost all massively multiplayer games or paper roleplay games is usually a hindrance . . . I thought it a good idea to try to curb this tendency.

(Gilles, developer)

15 Another compelling reason for giving preference to teams over individual players is linked to an innovative requirement the designers are very proud of, that is the possibility of playing the Jindeo game either from a PC (with extended game functionalities) or from a mobile terminal.

16 One Japanese lady player, interested in the collection of 'moments of time' completed her collection in the following way. She collected several moments of time icons one morning, and when back at home, she found a way to put them back on her virtual on-screen environment. Then she waited at home for the proper time to pick them up, thus subverting the principle of the game (which required her to move to definite places at given times to complete her collection).

17 There is a whole paper media industry dealing with this craze about starlet pictures, with fully dedicated tabloids.

18   To Goffman, cognitive recognition involves either placing another individual with respect to some singular information that characterizes him only, or with respect to 'some general social category, but in a context where any member of the category can play a crucial role' in the situation (as here the fact they are both logged on Nido players) (Goffman 1963).

19   Social recognition to Goffman is 'the process of openly welcoming or at last accepting the initiation of an engagement, as when a greeting or a smile is returned' (Goffman 1963).

20   Goffman himself emphasized that point, remarking that civil inattention was 'a courtesy that tends to treat those present merely as participants in the gathering and not in terms of other social characteristics' (Goffman 1963: 86).

21   The screen maps available to Nido users playing through their PC and the Internet server are much more extended than those available on mobile phone screen, which only figure the current cell.

Works Cited

Arendt, H (1983) *La Condition de l'homme*, Paris: Calmann Levy.

Benford, S, Anastasi, R, Flintham, M, Drozd, A, Crabtree, A, Greenhalgh, C, Tandavanitj, N, Adams, M and Row-Farr, J (2003) 'Coping with uncertainty in a location-based game', *Pervasive Computing*, 3: 34–41.

Castells, M (2001) *La Galaxie Internet*, Paris: Fayard.

Dodier, N (1995) *Les Hommes et les Machines*, Paris: Metailié.

Goffman, E (1963) *Behavior in Public Places*, New York: The Free Press.

Grint, K and Woolgar, S (1997) *The Machine at Work*, Cambridge: Polity.

Griswold, W, Shanahan, G, Brown, S, Boyer, R, Ratto, M, Shapiro, R and Truong, T (2004) 'ActiveCampus. Experiments in community-oriented ubiquitous computing' *IEEE Computer*, 37, 10: 77–81.

Hutchby, I (2000) *Conversation and Technology*, Cambridge: Polity.

Jones, Q, Sukeshini, G, Whittaker, S, Chivakula, K and Terveen, L (2004) 'Putting systems into place: a qualitative study of design requirements for location-aware community systems', in Proceedings of CSCW 2004, New York: ACM Press, pp. 202–211.

Kuhn, T (1962) *The Structure of Scientific Revolutions*, Chicago, IL: The University of Chicago Press.

Largier, A (2002) 'Je, nous, jeu. La constitution de collectifs de joueurs en réseau', *Réseaux*, 20, 114: 217–247.

Latour, B (1993) *La Clef de Berlin et autres Leçons d'un Amateur de Sciences*, Paris: Eds. La Découverte.

Lin, N (2001) *Social Capital: a theory of social structure and action*, Cambridge: Cambridge University Press.

Norman, D (1988) *The Psychology of Everyday Things*, New York: Basic Books.

—— (2003) 'Toilet paper algorithms: I didn't know you had to be a computer scientist to use toilet paper', http://www.jnd.org/dn.mss/ToiletPaperAlgorithms.html.

Quéré, L and Brezger, D (1992) 'L'étrangeté mutuelle des passants. Le mode de coexistence du public urbain', *Les Annales de la Recherche Urbaine*, 57–58: 88–99.

Rheingold, H (2000) *The Virtual Community. Homesteading on the Electronic Frontier* (revised edition), Cambridge, MA: MIT Press.

Sheller, M and Urry, J (2006) 'The new mobilities paradigm', *Environment and Planning A*, Special Issue on 'Materialities and Mobilities' (in press).

Shulga, T (2003) 'Présence médiatisée et construction de l'espace d'interaction: comparaison entre jeux de rôle classiques, MMORPG et jeux d'action en réseau', *Cahiers du numérique*, 4, 2: 101–115.

Simmel, G (1989) *Philosophie de la Modernité*, Paris: Payot.

Suchman, L (1987) *Plans and Situated Action: the problem of human-machine communication*, Cambridge: Cambridge University Press.

Thrift, N (2004) 'Remembering the technological unconscious by foregrounding knowledges of position', *Environment and Planning D: Society and Space*, 22: 175–190.

Turkle, S (1995) *Life on the Screen. Identity in the Age of the Internet*, New York: Touchstone.

Wittel, A (2001) 'Toward a network sociality', *Theory, Culture & Society*, 18, 6: 51–76.

CHAPTER TEN

Permeable Boundaries in the Software-sorted Society: Surveillance and Differentiations of Mobility[1]

David Murakami Wood and Stephen Graham

Values, opinions and rhetoric are frozen into code.
(Bowker and Star 1999: 35)

The modern city exists in a haze of software instructions. Nearly every urban practice is becoming mediated by code.
(Amin and Thrift 2001: 125)

Introduction

Differential mobility is in no way a new phenomenon; from the moment some people rode or were carried while others walked, there have existed differences in mobility which reflect and reinforce existing social structures. However, differential mobility is not always a matter of a simple correlation between greater wealth or status and greater speed; for example, in British cities, bicycle commuters tend to be better-educated with white-collar jobs. Mobility has always been configured by borders and boundaries composed of a multiplicity of hybrid objects, from infrastructure and technology to law and culture.

These boundaries are permeable to different degrees creating societies that are differentiated by speed and access, and the values attached to both. These values, in turn, reflect trends towards privatization, social polarization and the development of a risk-society, which might collectively be called neo-liberal or post-Fordist capitalism. That much is relatively uncontroversial: however, the role of technologies, those inhuman things, remains more problematic. What this chapter seeks to show is that there is a tendency towards technological lock-in, which threatens to divide contemporary societies more decisively into high-speed, high-mobility and connected and low-speed, low-mobility and disconnected, classes. The relative permeability of the boundaries that separate such groupings are increasingly orchestrated by automated systems of surveillance which continuously categorize and encode those categorizations and enforce them upon individuals or groups, based on the perceived danger that they are deemed to pose to the preferred social order of the 'kinetic elite' (Sloterdijk 1988). This is leading to an increasingly coded or software-sorted society and 'splintered' urban landscape characterized by highly differentiated mobilities: corridors of high mobility and easy access for some, and slow travel and difficult, expensive and blocked access for the majority. However, for neither class is the permeability particularly negotiable, whether or not such controls were originally accepted voluntarily or even requested (as most are in the case of higher income groups) or were enforced. Once introduced, both access and blockage are increasingly functions of encoded categorization. In this sense such a landscape of permeable boundaries is part of the broader edifice of 'new social control' identified by Michalis Lianos (2001, 2003), which threatens to remove the space of the city from everyday social construction.

In this chapter we explore the theoretical implications of these automated systems of surveillance, mobility control and boundary enforcement. We attempt to combine the work of Lianos on social control (which is still relatively unknown in the English-speaking world) with recent literature on code (Lessig 1999), the automatic production of space (Thrift and French 2002), and the actor-network-theory (hereafter, ANT) work of Latour, Callon, Law, Bowker and Star and others, which deals with the ordering processes of modern society. Our motivation for this approach is our belief that the work of actor-network theorists – as well as others working around the sociology of technology in areas such as Social Worlds Theory (Bowker and Star 1999) and Communities of Practice (Brown and Duguid 2000) – need to be far more integrated into mainstream social science understandings. This is especially necessary, we would argue, given the widening influence of attempts to reconstruct social theory using paradigms of flow, mobilities and process (for example, Urry 2000).

In all the enthusiasm for flows, it is crucial to remember that barriers and boundaries are as important as the flows themselves. In terms of structure, the chapter will first draw on ANT to outline the notion of the hybrid collective. The challenge of conceptualizing boundaries within this framework will then be addressed. We then move on to examine two recent approaches to the role of software in the construction of urban space and society – those of Lianos and Thrift and French – and discuss how their respective conclusions might be combined with an actor-network approach to boundaries, particularly with regard to systems of surveillance. Some examples are given, and we conclude by discussing how to establish the nature and location of power within a system of hybrid collectives which is both increasingly strongly ordered through technological systems and productive of class divisions. No firm answer is possible or perhaps even desirable, but we do give a number of alternative interpretations and trajectories. We emphasize, however, that this is not a technologically deter-minist vision, rather, it is a potential outcome of the extension and increasingly strong alignment of hybrid collectives, every component of which has human, nonhuman and inhuman elements mixed to some degree.

Hybrid Collectives

Deriving from the philosophy of Michel Serres and Michel Foucault, and originating in the sociology of science and technology, ANT is an evolving theoretical and methodological grounding for the social sciences, which seeks a middle ground between the extremes of structuralism and agency-centred approaches such as ethnomethodology and microsociology. Some have gone so far as to claim that ANT seeks to break down all the dualisms that are felt to afflict the study of society, for example: global/local, social/natural, tech-nical/social etc. Despite its origins, empirical and theoretical work based on or adapting ANT has proliferated and infiltrated other disciplines.[2]

ANT attempts to derive explanations for two fundamental problems in the study of societies: first, the way in which long-lasting social structures appear out of social interactions; and, second, the method by which power can act at a distance. Actor-network theorists argue that the key element in both these processes is the interaction between humans, nonhumans (other living beings, natural processes etc.) and inhumans (man-made objects, materials, texts and so forth). These nonhumans and inhumans are not merely passive in the sense of being things imbued with value by society; rather, they carry, change and produce power and value in a symmetrical relationship with individuals and groups of human beings. Thus, human societies do not exist solely by the inter-actions of individuals, but only because of the crucial role of nonhumans and inhumans in making these interactions last beyond their specific occurrence in time and space. This, in essence, is the origin of social power: it is always an outcome of interactions between actants, neither a property of any one actant, nor an abstract force.

ANT therefore accepts a performative conception of social relations but extends it (Latour 1986): ANT argues that, through their *programmes of action* (from innate tendencies to volition), *actants* attempt to enrol other actants to fulfil those programmes, producing hybrids. These relationships develop through the passing of intermediaries between actants. In this way actants perform, and are performed by, complex, shifting sociotechnical networks, which produce space, identity and social relations (Michael 1996; Murdoch 1997a, 1997b). Importantly, *actor-networks*, or *collectives*, are neither amorphous nor uniform. Collectives can be variably stable over time (irreversibility). They can contain actants whose programmes of action are more or less strongly aligned (conver-gence). And they can be variably bounded or territorial (Callon 1991). It is to this last concept of boundaries that we shall now turn as it is key to notions of differentiated mobility within societies orchestrated by code.

Categorization and Boundaries

This chapter adopts a philosophical base from surveillance theory (see Lyon 2003) and which has been explored further elsewhere by one of the authors (Donaldson and Wood 2004). This is that processes of categorization or classification are central to both the practice and understanding of social control. This notion of social ordering as control is also present implicitly within ANT-type approaches: John Law (1992) has defined 'ordering' as a key characteristic of modern society and Bowker and Star (1999) argue that modernization has created more and more detailed and differentiated, and increasingly enforced or policed, classifications. These categories can only function because of boundaries that delineate one category from another.

Within the study of technologies, Michel Callon has reintroduced the notion of the boundary (Callon 1991).[3] In Callon's variation on ANT, which refers to techno-economic networks (TENs), boundaries of TENs relate to the level of convergence within a TEN: 'an element may be treated as lying outside a network if it weakens the alignment and coordination – that is the convergence – of the latter when moved into the network' (p. 148). Convergence 'measures the extent to which the process of translation and its circulation of intermediaries leads to agreement' (p. 144).

Agreement consists of two dimensions: *alignment* and *co-ordination*. Alignment is the extent to which translation '*generates* a shared space, equivalence and commensurability' (p. 145), its opposite being disalignment. Between these two extremes are various degrees or strengths of alignment. In any process of translation, the imputation of authorship is crucial; this is determined by rules about: the identity of actors, the imputation of intermediaries to actors; and finally 'who may *speak* on behalf of whom' (p. 146). Co-ordination refers to the process of 'organising imputation and limiting the number of transactions that can be easily stabilized' (p. 147). Codified rules are called *co-ordination* (or *translation*) *regimes*. *Weak co-ordination* refers to general networks with no specifically local rules, whereas *strong co-ordination* refers to a network with strong local and general rules. The stronger the co-ordination, the more predictable the network. Convergence, then, is the degree to which a combination of alignment and co-ordination is in effect.

What is it that these TENs, or hybrid collectives, attempt to do? Akrich and Latour describe this as the production of meaning or 'how one privileged trajectory is built, out of an indefinite number of possibilities' (Akrich and Latour 1992: 259). John Law has elsewhere described this as the way in which a social mode of ordering is produced (Law 1992). Bowker and Star (1999) call it 'categorical work' – the creation of static categories from a fluid reality. In essence, then, hybrid collectives attempt to define and/or further expand their own boundaries through the hiding or erasure of alternate possibilities through categorical work, that is, the creation of both discursive and material boundaries. It is to the way in which these boundary-making processes affect the space of the city that we now turn.

New Social Control in the Software-Sorted City

Within the city (or indeed any area) boundaries, their permeability (what is allowed to pass through any boundary and how), and the nature of the spaces separated by the boundary, are based on prior categorical work. This is accomplished and enforced through surveillance, which can therefore be conceptualized as a social mode of ordering, in the sense that John Law uses the term. Fundamentally, surveillance is not about individual people, but about defining the relationship of all sorts of actants in relation to boundaries: it is a technocratic form of territoriality and the (attempted) control of mobilities and flows. It is the determination of particular spaces and relationships to those spaces through categorization, boundary maintenance (in terms of both space and identity), observation and enforcement. Surveillance, in its most extremely territorialized manifestations, depends therefore on the purity of categories, and on the cleanest possible demarcations between them, what Graham refers to in passing as 'the precision of . . . boundary maintenance' (1998: 449). Without boundaries, categories dissolve, and the enforcement of territoriality through surveillance becomes impossible.

However, the most subtle and successful boundaries are *im*precise. More accurately they allow different speeds of movement and different actants to move in variable ways and speeds in relation to the boundary. They are therefore *differentially* permeable. Turning such a conceptualization on its side and looking at it from the point of view of mobilities rather than territory, one can see a series of different geographies of mobility overlaid like a palimpsest with territoriality cutting through them in jagged and uneven ways.

Crucially, both the creation of differentially permeable boundaries in cities and their surveillant enforcement are increasingly carried out by computer software. Several theorists have argued that the interactions between systems based on code or software, and the humans subject to software-sorting as they move around the city (or the world), produces a dehumanizing or desocializing process. A paper by Nigel Thrift and Shaun French, for example, argues that the pervasiveness of automated software-driven systems in contemporary Euro-American societies means that it is now possible to speak of 'the automatic production of space' (Thrift and French 2002).

At the same time, the French sociologist of surveillance Michalis Lianos has been developing a new theory of social control to address the recognition that technologies are basically about spatial control. As he puts it, 'a technological system is by definition a system of control of a certain environment' (Lianos 2003: 418). Automation is particularly important within Lianos' thinking because such an automated surveillance system 'transforms the spatial threshold of the institution into a threshold of legitimacy' (ibid.: 420). In other words, the legitimacy of an action in relation to the automated systems is removed from the sphere of social negotiation. As Lianos and Douglas previously noted, such systems – which they term 'Automated Socio-technical Environments' (ASTEs) – *'radically transform the cultural register of the societies in which they operate by introducing non-negotiable contexts of inter-action'* (Lianos and Douglas 2000: 107, italics in original). Automated surveillance systems, based on computer code, thereby accelerate the trend away from persons towards data images as the basis for trust in society (see Lyon 2001, ch. 5). Automating surveillance can thus potentially mean the automation of social representation.

Lianos argues that a relational perspective is essential to understand such a transition. This is because:

> *the criterion for deciding what belongs or not to the sphere of control is neither the consciousness of the subject or the group involved, nor the will of those who produce the 'controlling' effect in question, but mainly the conditions that shape the interaction between those two parties.*
>
> (Lianos 2003: 416, italics in original)

Indeed, Lianos continues that, in most cases, social control is not the intended effect of the collectives[4] producing it. To him, 'it is . . . necessary to recognise the existence of types of activity where control arises in many ways that were often not intended to produce a controlling effect' (ibid.).

For Lianos, then, what is occurring is not a deliberate form of oppressive control but an institutional – bureaucratic obsession with function, with the smooth flow of goods and services, and with efficiencies of movement and transactional fluidity:

> What the subject thinks, does or believes, is irrelevant to what the institution controls; it is simply meaningless for the technological device. In the thousands of daily transactions with institutional outlets, which weave together the sociocultural resources of the postmodern subject, there is not a single wish to build and promote a cognitive and moral universe. The exclusive objective that is being pursued is to stimulate and facilitate behaviours that are favourable to the effective functioning of organisations.
>
> (Lianos 2003: 423)

What is really going on here, to Lianos, is a decline of collective sociality (in whatever form) and the creation of a new form of institutional sociality, increasingly governed by inhuman rules of automated flow which are orchestrated and enacted through enormous systems of interlinked and computerized elements using code. Thus, in a sense, the mobility of things is privileged ahead of the mobility of persons, or more accurately, certain categories of people. This proliferation of automatic systems generates, as Norris *et al.* note, a concern about exclusion: 'the problem with automated systems is that they aim to facilitate exclusionary rather than inclusionary goals' (Norris *et al.* 1998: 271). As Norris has written more recently, 'it is not the inclusionary project envisaged in Bentham's panopticon that will become operationalised by the spread of digitalised enforcement, but exclusion' (Norris 2003).

There is a potential problem here with our earlier advocacy of power as an outcome of interactions between hybrid collectives. Lianos does not use the word 'collectives' and appears to dismiss ANT and its variations on the grounds that 'the central question remains that of the "integration" of technology in the context of the distinction between nature/culture/society, rather than the analysis of its content' (Lianos 2003: 419). However, we believe this to be a misreading or at least a failure to recognize the potential of ANT. What Lianos is describing in the quotation above is, in Latourian terms, the 'program of action' of technological devices concerned. Such devices have no ideational content in themselves. However, these devices themselves are not and cannot be isolated: they form part of larger hybrid collectives. As Lianos himself describes, they carry out functions that reflect the programme of action of very strongly aligned and strongly co-ordinated collectives (which he calls organizations).

Such collectives exist within the very large, strongly aligned although less strongly co-ordinated collective of neo-liberal capitalism, which, if nothing else, is the program of action of certain prioritized and privileged actants, translated through practices, laws, and technologies into an increasingly global set of technosocial collectives. That some of the effects of particular technologies are unintentional as far as any organization is concerned is unsurprising – no process of translation is ever complete. But, insofar as the programmes of action of technologies do nothing but regulate the flow of the objects and peoples as the programme of action of the organization intends, why should the organization object?

The Automatic Sorting of Mobilities: Some Examples

There are several examples of how these hybrid collectives create permeable boundaries that are orchestrated to produce vastly differential mobilities. One is electronic road pricing, where new, heavily monitored private premium highways and road spaces are created that are only accessible to drivers with in-car electronic transponders. Access to road space thus becomes a priced commodity dependent on users having the technology in their cars, and the resources (and often bank accounts) to pay bills for a newly commodified good that cannot be bargained for – you either have the intermediary that allows you to pass the automated system or you do not. This reengineering of public roadspace monopolies into separated domains of premium, charged highway space, and uncharged highway space that continues to be freely accessible, is orchestrated and policed through vast hybrid collectives based on code (call centres, tracking cameras, financial management systems, geodemographic marketing systems, enforcement and prosecution systems, and so on).

A second example of the use of code to orchestrate permeable mobilities across boundaries is drawn from international airports. There are several ways in which surveillance technologies, from the Immigration and Naturalization Service Passenger Accelerated Service System (INSPASS), which uses hand-geometry – a relatively old and unsophisticated system – to the highly advanced iris-scanning programs operated by Amsterdam's Schipol airport since 2001 and now required by US Naturalization and Immigration services. Such systems effectively automate normative judgements encoded into immigration control databases into boundary-maintenance: frequent travellers and business flyers are 'trusted' and automatically allowed to pass, if their biometric signature matches that on a card or in a database. At the same time, all others become suspect or are even automatically deemed 'treacherous' because they are not warranted entry into the speeded-up world of elite immigration control which bypasses conventional borders just as the speeded-up flows of computerized and priced highways bypass the world of the public highway system (Curry 2003). National boundaries are thus differentially permeable boundaries increasingly policed by automated systems. Indeed, this challenges the whole political concept of 'national borders' for some, while making such borders more rigid and impenetrable for others. The 'kinetic elites' discussed by Sloterdijk are enrolled into hybrid collectives that allow them to transgress national borders. At the same time, new surveillance systems are installed so that 'illegal' migrants and refugees find that such borders become more fortress-like (see, for example, Verstraete 2001; Flynn 2003).

A final pair of examples are more invisible. First, Internet routers are now being programmed with software that actually prioritizes packets of information differently, based on real-time, corporate judgements of the real or potential profitability of the person sending the traffic. This means that the 'premium' users seen to be more commercially attractive may, over the same infrastructure, be able to access web sites in times of Internet congestion while other, less attractive users experience 'web site not available' signs. As a result, corporate Internet firms now routinely prioritize the traffic from the 'premium', selected users that they think will bring the best revenues and exposure to their brands, their services, and their corporate tie-ins. At the same time, they actually downgrade the priority of web and Internet traffic that is deemed to be of marginal profit potential or that which will benefit competing firms – what they often term the 'scavenger' class (Graham 2005). Importantly – and in sharp distinction to the visibility of differentiation in the highways or airports examples – neither the privileged nor the marginalized users in this case are likely to even know that this process has shaped their experience of the network.

In a detailed analysis of how growing corporate control of the Internet is configuring its media spaces, Dwayne Winseck (2003) found that such trends towards the software-sorting of Internet users were highly advanced. In 1999, for example, Cisco, the manufacturer of most of the Internet's routers, advertised to corporate media and Internet firms offering them:

> absolute control, down to the packet, in your hands. . . . You can identify each traffic type – Web, e-mail, voice, video . . . [and] isolate . . . the

type of application, even down to *specific brands*, by the *interface used*, by the *user type and individual user identification or by the site address*.
(cited in Winsbeck 2003: 183, original emphasis)

Given that the Internet was purposefully constructed so that each 'packet' of data had equal priority – a system termed the 'best effort' Internet – such 'unbundlings' or 'splinterings' of users' experiences, based on corporate control and software-sorting, potentially has profound political and social implications.

In our final example, an equally invisible process of differentiated mobilities is emerging in call centre queuing systems. Here, automated systems are rapidly emerging. These can queue incoming calls differently, depending on inbuilt, algorithmic, judgements of the profits the company makes from them. 'Good' customers can thus be answered quickly while 'bad' ones are put on hold. As with Internet prioritization, neither user is likely to know that such prioritization and distancing is occurring. It is worth pointing out that here, as elsewhere, one could argue that the 'control' of customers is an entirely irrelevant part of the system as far as the sorting devices or, indeed, the companies using the software are concerned: their programme of action is to ensure the maximization of profits – 'good' in this sense equates to high or prompt-paying – but differentiation and controlling effects are an inevitable outcome.

The Politics of Software-sorted Societies

Automated systems quite clearly enforce (and increasingly create) categories and boundaries. This might sound like the kind of technologically determinist perspective that is often found in writing on ICTs and which relational theories seek to overcome (see Bingham 1996). Thrift and French set out two main reasons why grand visions of automatic cities are overstating the case. First, they argue that software is still designed and initiated by human beings, and are often written with 'human' concerns in mind, through a series of virtual moments of mobilization. Second, they suggest that ethnographic understandings of the function of software are necessary which tend to undermine notions of vast, automated, and coded systems orchestrating urban life. Drawing on the work of Andrew Barry, Thrift and French argue that this growth of coded systems of automated flow, mobility and surveillance presents a challenge to politics – the axiom that all politics is effectively now the politics of technology (Barry 2001). The human is thus not eliminated but simply more hidden. Through the use of automated systems of surveillance and control, the worldviews, social judgements, and ordering processes that are inseparable from processes of inventing and enforcing boundaries, can be hidden. As ANT would have it, such systems can become black-boxed and punctualized such that the collective itself becomes a seemingly integrated and individual component in broader collectives. In other words, the vast technosocial collectives that constitute boundaries through enacting differential mobilities automatically emerge to become the very taken for granted and ordinary landscapes of contemporary cities and systems of cities. Such processes of black-boxing do not happen through technological determinism. Rather, processes of techno-logical lock-in of punctualized collectives occur that are utterly social, often accidental, and infused with human and institutional agency and the politics of code. In examining any technology, one finds that what is actually doing the work is a combination of technological, human and natural functioning in some collective form.

However, there is a very serious sociopolitical problem here: this process of 'looking', of examination, is itself affected both by the discursive and mater-ial boundaries created by the automation of such systems, and the languages and expertise required to understand their functioning. Conventional surveil-lance systems already influence mobilities within, through and between the public and private spaces of urban everyday life. However, their impact is moder-ated by the three-fold interaction between the subjects, the technologies and their human operators. With the rise of software-sorting systems that differen-tiate mobilities across borders continuously, automatically, and in real-time, however, there is an overt attempt to configure human behaviours remotely, whether that is for malign, beneficial or neutral intentions, without the uncer-tainty provided by direct human operation. The potential for discrimination, and indeed for transgression, has effectively been given to the designers, builders and programmers of such systems, who are able to embody their prejudices and desires into the very functioning code and architecture of the systems them-selves. Such systems are often promoted as infallible, logical and free of human prejudice. In fact, many programmers do not understand that they are in any way responsible for the 'unintended' effects of discriminatory systems, nor even that the systems themselves can be considered 'discriminatory'.

The diffusion of these automated hybrid collectives are thus changing the very nature of space, and the balance between deterritorialization and re-territorialization that shapes contemporary urban life. Public spaces have long been the arenas for popular protest and politics at its most raw and direct. The politics of human collective mobility and gathering are, in many cases, being displaced into the more obscure collectives of the technologies themselves, the architecture, and the code that drives them. For politics, which has been internalized, there is still territoriality, but it relates not just to the 'outside', to mobility in material space, but also to the 'inside', the virtual space of flows of databases and networks. They are perhaps better understood as topologies rather than spaces in the usual sense. Topology has two relevant scientific meanings here: in computer technology, it refers to the physical pattern of connectivity within a computer network or between processors, memories, and peripherals (Harcourt n.d.); and in mathematics, topology is the study of prop-erties of objects that are preserved through deformations – it is the study of spatial objects (Weisstein n.d.). This is the second sense in which boundaries are differentially permeable, beyond the theme of this chapter, in that they flow in and out of and between virtual and material space, changing as they go.

Many of these interior spaces, these topologies, are intrinsically territorial: intranets are controlled through physical barriers as well as informational ones (passwords, firewalls, etc.); databases may exist on computers that have no external connectivity beyond screen, drive and keyboard.[5] However, there are also more subtle forms of boundary-marking, which are vital to an understanding of surveillance. These boundaries are largely socio-cultural and discursive, produced by such entities as 'communities of practice' (Star 1995; Wenger 1998); they are networks formed through learning and work-based activity, or 'epistemic communities'. Star (1995) shows that it makes sense to portray computer-based social groups as communities of practice, both in terms of the production of situated knowledge and the technologies themselves, and the experience of the users of those systems. In the case of software-driven surveillance, the direct users are other computer systems (though of course the ultimate although indirect users are human), but our attention is drawn immediately to the producers of the systems (the engineers). These producers are often software engineers, or research groups that include software, hardware and network engineers. These collectives are relatively opaque to outsiders. Their arguments do not spill out much beyond particular Internet webzines or books only likely to be read by the already computer literate. In territorial terms, there are some physical barriers (lack of access to appropriate computer hardware), but the boundaries are created more by assumptions of knowledge, interests and other practice-based qualifications. Thus, while the programming communities may be committed to freedom of information, the methods by which they distribute information are intensely circumscribed.

Conclusions and Future Directions

While the 'spaces of flows' (Castells 1996) are increasingly being automated, they are also increasingly complex. Surely this poses immense problems for such systems. How can software-sorting, based on categorical judgements and differentiated mobilities, cope with the levels of complexity and the multiple intermediaries that exist? A first option here, of course, is to restructure society so that it is amenable to control. As Lianos suggests, the roll-out of smart ID cards to replace all other forms of identification and trust-substituting intermediaries is but one example. This, of course, requires the integration of databases of all kinds at the highest level at which such information is required – effectively and eventually a global level. In a space of global flows one would not expect national biometric identity cards to remain purely national for very long. Of course, there are problems with various particular surveillance technologies in particular settings – facial recognition does not seem to work very well in public space yet, and may in fact have fundamental flaws as a concept (see Introna and Wood 2004). However, this simply means that other systems – in this case, for example, the automated recognition of human walking styles or gait – will be attempted, instead.

187

A second approach is to attempt to improve the surveillance systems to cope with multiple ordering, and with the vast complexities of flow across boundaries, to make judgements beyond the binary, in other words to move towards heuristic surveillance. Heuristic systems are learning systems. That is, they are not based on simple algorithms, but have algorithms that enable the writing of additional new algorithms to cope with new knowledge and new situations. The result is self-programming software.

At present, one thing is clear. Within the context of the widespread privatization of urban and mobility spaces across the world, software-sorting techniques are being socially shaped in two very different ways. On the one hand, the surveillance and monitoring capacities of ICTs are being shaped to prioritize and enhance the power and mobilities of privileged human bodies within the many scales of global, neo-liberal capitalism. On the other, ICTs are being configured to *add* friction, barriers or logistical costs to the mobility and everyday lives of those deemed by dominant states or service providers to be risky, unprofitable, or undeserving of mobility.

Crucially, these techniques of prioritization and inhibition are often so invisible and automated that neither the losers nor the beneficiaries are even aware that they are in operation within the complex sociotechnologies that increasingly constitute the ordinary and taken-for-granted environments of contemporary societies (Graham and Wood 2003). Thus, the broadly standardized infrastructures and mobility services and rights to public space that were a key part of the elaboration of Keynesian and Fordist welfare states in many nations are often being replaced or overlaid with consumerist systems through which individualized and market-based entitlements to the rights of mobility and of the city are enacted. Automated software increasingly polices these judgements as cross-subsidies are abandoned between profitable and unprofitable users and areas in the desire to extract profits only from those who can provide it.

Importantly, the use of software-sorting to 'unbundled' public spaces and Keynesian, welfarist or redistributive service regimes and domains is extending rapidly. Public spaces and malls in the UK and US are being sorted through face recognition CCTV. Mutual health insurance threatens to be unbundled through individualized assessments of genetic risk based on genomic research. And even retailers have debated the use of software-sorting. In 1999 UK supermarkets debated using ICTs to bring in a two-tier pricing policy which would use high prices at peak times to shift the cash-poor time-rich citizens of cities out of the way of cash-rich/time-poor ones (an idea that was then abandoned).

As a result of these processes of change, the clear risk is that the previously largely integrated public domains, infrastructures and spaces of cities will become progressively 'splintered' in socio-technical terms (Graham and Marvin 2001). Rapidly emerging are separated socio-technical realms of premium and marginalized citizens that are structured so that citizens experience totally different life chances, access rights, and service qualities based on the continuous and automated judgements of vast arrays of (often unknowable) software orchestrating and enacting myriad boundaries through differentiated mobilities.

While sometimes these splintering domains are clearly visible, more often they are totally invisible. Their agency can even be missed by the people who are sifted by them – either the winners or the losers. Increasingly, then, the politics of the right to the city amounts to the hidden politics of code as the agency of software structures urban access and exclusion in subtle but powerful ways (Thrift and French 2002).

Notes

1 Stephen Graham would like to thank the British Academy for the support of a Research Readership (2003–2005) without which his contribution to this research would not have been possible.
2 There have of course been many other attempts to create integrated process-based social theory, most notably Gidden's Structuration Theory (Giddens 1984), the revival of dialectics, attempts to integrate complexity theory into social science, and the 'new institutionalism' (particularly historical institutionalism) in politics and organization theory.
3 The work of Barth (1969, 2000) and Cohen (2000) on boundaries has been vital.
4 Lianos, of course, does not use this term.
5 This, however, does not mean that such databases are uncrackable. Van Eck phreaking, the ability to capture and read the radio emanations from CRT computer screens and other electromagnetic devices, has been known for some time; and it has recently been proposed that covert optical surveillance of light-emitting computer devices may be possible. Thus databases in use can be read remotely from the screens of the user.

Bibliography

Akrich, M and Latour, B (1992) 'A summary of a convenient vocabulary for the semiotics of human and non-human assemblages', in W Bijker and J Law (eds) *Shaping Technology/Building Society*, Cambridge, MA: MIT Press.
Amin, A and Thrift, N (2001) *Cities: reimagining the urban*, Cambridge: Polity.
Barry, A (2001) *Political Machines*, London: Athlone.
Barth, F (ed.) (1969) *Ethnic Groups and Boundaries*, London: Allen & Unwin.
—— (2000) 'Boundaries and connections', in A P Cohen (ed.) *Signifying Identities: anthropological perspectives on boundaries and contested values*, London: Routledge.
Bingham, N (1996) 'Object-ions: from technological determinism towards geographies of relations', *Environment and Planning D: Society and Space*, 14: 635–657.
Bowker, G and Star, S (1999) *Sorting Things Out*, Cambridge, MA: MIT Press.
Brown, J-S and Duguid, P (2000) *The Social Life of Information*, Cambridge, MA: HBS Press.
Callon, M (1991) 'Techno-economic networks and irreversibility', in J Law (ed.) *A Sociology of Monsters: essays on power, technology and domination, Sociological Review Monograph 38*, London: Routledge.
—— (1998) 'An essay on framing and overflowing: economic externalities revisited by sociology', in M Callon (ed.) *The Laws of the Markets*, Oxford: Blackwell/*The Sociological Review*, 244–269.
—— and Law, J (1995) 'Agency and the hybrid collectif', *South Atlantic Quarterly*, 94: 481–507.
Castells, M (1996) *The Rise of the Network Society, the Information Age: economy, society and culture*, Oxford: Blackwell.

Cohen, A P (ed.) (2000) *Signifying Identities: anthropological perspectives on boundaries and contested values*, London: Routledge.

Curry, M R (2003) 'The profiler's question and the treacherous traveler: narratives of belonging in commercial aviation' *Surveillance & Society*, 1 (4): 475–499. Online. Available http://www.surveillance-and-society.org/articles1(4)/treacherous.pdf (accessed 31 January 2005).

Donaldson, A and Wood, D (2004) 'Surveilling strange materialities: the evolving geographies of FMD biosecurity in the UK', *Environment and Planning D: Society and Space*, 22 (3): 373–391.

Flynn, S E (2003) 'The false conundrum: continental integration versus homeland security', in P Andreas and T J Biersteker (eds) *The Rebordering of North America: integration and exclusion in a new security context*, New York: Routledge.

Foucault, M (1975) *Discipline and Punish: the birth of the prison*, New York: Vintage.

Giddens, A (1984) *The Constitution of Society: outline of the theory of structuration*, Cambridge: Polity Press.

Graham, S (1998) 'Spaces of surveillant-simulation: new technologies, digital representations, and material geographies', *Environment and Planning D: Society and Space*, 16: 483–504.

—— (2005) 'Software-sorted geographies', *Progress in Human Geography* 29 (5): 1–19.

—— and Marvin, S (2001) *Splintering Urbanism: networked infrastructure, technological mobilities and the urban condition*, London: Routledge.

—— and Wood, D (2003) 'Digitising surveillance: categorisation, space and inequality', *Critical Social Policy* 23 (2): 227–243.

Harcourt Academic Press Dictionary of Science and Technology, Online. Available http://www.harcourt.com/dictionary/browse/19/ (accessed 31 January 2005).

Introna, L and Wood, D (2004) 'Picturing algorithmic surveillance: the politics of facial recognition systems', *Surveillance & Society*, 2 (2/3): 177–198. Online. Available http://www.surveillance-and-society.org/articles2(2)/algorithmic.pdf (accessed 31 January 2005).

Jones, R (2000) 'Digital rule: punishment, control and technology', *Punishment and Society*, 2 (1): 23–39.

Latour, B (1986) 'The powers of association', in J Law (ed.) *Power, Action and Belief: a new sociology of knowledge? Sociological Review Monograph 32*, London: Routledge and Kegan Paul.

—— (1991) 'Technology is society made durable', in J Law (ed.) *A Sociology of Monsters: essays on power, technology and domination, Sociological Review Monograph 38*, London: Routledge.

Law, J (1992) 'Notes on the theory of the actor network: ordering, strategy and heterogeneity', Online. Available http://www.lancs.ac.uk/fss/sociology/papers/law-notes-on-ant.pdf (accessed 31 January 2005).

—— (1999) 'After ANT: complexity, naming and topology', in J Law and J Hassard (eds), *Actor-Network Theory and After*, Oxford: Blackwell/The Sociological Review, 1–14.

Lessig, L (1999) *Code*, New York: Basic Books.

Lianos, M (2001) *Le Nouveau Contrôle Social*, Paris: L'Harmattan.

—— (2003) 'Social control after Foucault', *Surveillance & Society*, 1 (3): 412–430. Online. Available http://www.surveillance-and-society.org/articles1(3)/AfterFoucault.pdf (accessed 31 January 2005).

—— and Douglas, M (2000) 'Dangerization and the end of deviance', *British Journal of Criminology*, 40: 261–278.

Loader, B (ed.) (1997) *Cyberspace Divides*, London: Routledge.

Lyon, D (2001) *Surveillance Society*, Buckingham: Open University Press.

—— (ed.) (2003) *Surveillance as Social Sorting: privacy, risk and automated discrimintion*, London: Routledge.

Michael, M (1996) *Constructing Identity: the social, the nonhuman and change*, London: Sage.

Murdoch, J (1997a) 'Towards a geography of heterogeneous associations', *Progress in Human Geography*, 21 (3): 321–337.
—— (1997b) 'Inhuman/nonhuman/human: actor-network theory and the prospects for a nondualistic and symmetrical perspective on nature and society', *Environment and Planning D: Society and Space*, 15: 731–756.
Negroponte, N (1995) *Being Digital*, London: Hodder & Stoughton.
Norris, C (2003) 'From personal to digital: CCTV, the Panopticon and the technological mediation of suspicion and social control', in D Lyon (ed.) *Surveillance as Social Sorting: privacy, risk and automated discrimination*, London: Routledge, pp. 249–281.
——, Moran, J and Armstrong, G (1998) 'Algorithmic surveillance: the future of automated visual surveillance', in C Morris, J Moran and G Armstrong (eds) *Surveillance, Closed Circuit Television and Social Control*, Aldershot: Ashgate, pp. 255–267.
Serres, M (1974) *La Traduction, Hermès III*, Paris: Editions de Minuit.
Sloterdijk, P (1988) *Critique of Cynical Reason*, London: Verso.
Social Exclusion Unit (SEU) (2000) *Bridging the Digital Divide*, London: DTI. Online. Available http://www.cabinet-office.gov.uk/seu/2000/pat15.doc (accessed 31 January 2005).
Speak, S and Graham, S (1999) 'Service not included: marginalised neighbourhoods, private service disinvestment, and compound social exclusion', *Environment and Planning A*, 31: 1985–2001.
Star, S L (1995) 'Introduction', in S L Star (ed.) *The Cultures of Computing*, Oxford: Blackwell/*The Sociological Review*, pp. 1–28.
Thrift, N and French, S (2002) 'The Automatic Production of Space', *Transactions of the Institute of British Geographers*, 27 (4): 309–335.
Tsung Leong, S (2001) 'Ulterior spaces', in C Chung, J Inaba, R Koolhaas and S Tsung Leong (eds) *Harvard Design School Guide to Shopping*, Taschen: Cologne, pp. 765–774.
Urry, J (2000) *Sociology Beyond Societies: mobilities for the twenty-first century*, London: Routledge.
Utility Week (1995) *Special Issue: IT in Utilities*, 19 November.
Verstraete, G (2001) 'Technological frontiers and the politics of mobility in the European Union', *New Formations*, 43: 36–43.
Virilio, P (1991) *The Lost Dimension*, New York: Semiotext(e).
Weisstein, E (n.d.) '"Topology", Eric Weisstein's World of Mathematics. Online. Available http://mathworld.wolfram.com/Topology.html (accessed 31 January 2005).
Wenger, E. (1998) *Communities of Practice: learning, meaning and identity (Learning in Doing: social, cognitive and computational perspectives)*, Cambridge: Cambridge University Press.
Winseck, D (2003) 'Netscapes of power: convergence, network design, walled gardens, and other strategies of control in the information age', in D Lyon (ed.) *Surveillance as Social Sorting: privacy, risk and automated discrimination*, London: Routledge, pp. 176–198.
Young, J (1999) *The Exclusive Society*, London: Sage.

INDEX

Page references to notes are followed by the letter 'n', while references to non-textual matter such as Tables are in *italic* print

accessibility: and get-at-ability 103, 104–5; lack of, and social exclusion 103–4; poverty of 104; space-time theory 104, 115n
accessibility diaries 12, 110–13, *111–12*, 115, 116n
accumulation paradigm, design process 167
actants 179
activity codes, diaries 116–17n
actor networks 179
actor-network theory (ANT) 46, 178, 179, 183, 185; surveillance 14; theoretic tools inspired by 13
Adey, Peter 11, 44–60
Afghanistan: cruise missile attacks 122; western occupation 130
Africa, sub-Saharan, colonial expansion 123
agreement, dimensions of 180
AI *see* artificial intelligence
Air Traffic control system (US) 129
airlines, no-frills 46
airport city, physical space of 52
Airport Management Operational Support System (AMOSS) 53
airports, international 11; and code, use of 52, 183–4; technology 14
air travel, terrorist fears 13
Aitchison, C. 25
Akrich, M. 180
alignment 180
al Jazeera, news network 129
Alternative Mobility Futures conference, Lancaster University (Centre for Mobilities Research) 15, 57n

Amin, A. 10, 11
AMOSS *see* Airport Management Operational Support System
Amoy Gardens 131n
Amsterdam, Schipol airport 184
APEC *see* Asia Pacific Economic Cooperation
Appadurai, Arjun 8–9, 10, 143
architecture: ciudad lineal movement 22; recombinant 47
Arendt, Hannah 168
Armitage, J. 4
artificial intelligence 155; 1980s projects 129
Asia Pacific Economic Cooperation (APEC) 128
Association of Train Operating Companies (ATOC) 79, 87, 88, 90
ASTEs *see* Automated Socio-technical Environments
ATOC *see* Association of Train Operating Companies
Attali, J. 127
Augé, Marc 51, 70, 71
Automated Socio-technical Environments (ASTEs) 181–2
automation 181
automobility: rise in twentieth century 83; and *Star* film 68, 69, 71

Ballard, J. G. 69
Barrell, J. 25
Barry, Andrew 185
Barton, William 30

Batty, M. 105
Beckman, Karen 68, 69
Benjamin, Walter 11, 69
Bevan, Paul 11, 44–60
bio-terrorism 129
black-boxing processes 185–6
Blade Runner 10, 159
BMW car: and corporate interests 62; in
    film 61–2, 64, 65, 66; 5-series model
    67–8; 'pleasure of driving' logo 67;
    safety engineering 68; as symbol of
    1990s dot.com wealth 72
Bond, James 65, 72
boundaries and borders 14, 177–8; and
    categorization 180; and examination 186;
    imprecise 181; permeable 181
boundary-marking 187
Bowker, G. 178, 180
'Bradshaw's Railway Guide' 86, 91
Breitling, P. 23
Brenner, P. J. 40n
Broadreach (Internet Service Provider) 138,
    139, 145
Bruntland Report, World Commission on
    Environment and Development (1987)
    126
business class mobility 12
Busson, Paul 38
bypassing: blockages and disruption 92–5;
    implications for public transport
    providers 97–8; notion of 81, 97

call centre queuing systems 185
Callon, Michel 142, 143, 147, 178, 180
capitalism, neo-liberal 183
'Capitalist Realism' 74n
car as mobile commodity 10, 62, 63
car culture films, twentieth-century 12,
    61–74
Carter, President Jimmy 63
case studies 3
Castells, Manuel 149
categorical work 180
categorization, and boundaries 180
celebrities, in film *see* Madonna; Owen,
    Clive; Ritchie, Guy
cell phones *see* mobile phones
Center for Disease Control (CDC), Atlanta
    128
Changi airport 52
Chicago School urban theorists 1
China, SARS in 123

Church of St Augustine (Vienna) 30–1
Cisco (manufacturer of Internet routers)
    184–5
cities: disease movements 2; information
    flows 1–2; 'metabolism' 1; migrants 2;
    and mobility 1–17, 46–8; social control
    181–3; socio-technical side of 46;
    terrorism fears 2; tourism and travel 2;
    24-hour, concept 96; 'undiscovered', and
    linearity 22
city flows, concept 103
city guides 37
City Link, Melbourne 47
ciudad lineal movement, architecture 22
civil inattention concept 173; equivalence
    principle 172
Clarke, R. 129
clock time 81, 85
clustered exclusion, in cities 103
clusters 7
code, and airports 183–4
Cold War, bipolar discipline of 126–7
collectives: hybrid 179, 180, 183, 186;
    technosocial 185–6
communication, human-machine 91
communication paradigm/affordances 154
communication technologies 4; and physical
    mobility 12
Communities of Practice 178
complaints, transport 98n
complementarity, physical and virtual
    mobility 108
complexity: dimension of 125; and risk
    124–6
Computer Professionals for Social
    Responsibility (CPSR) 128
Consume project (Greenwich) 147
convergence 180
co-ordination 180
co-presence 7; defined 12; and ICTs 84;
    synchronizing 81–5; and technology
    122
Couclelis, H. 106
coupling, dimension of 125
CPSR *see* Computer Professionals for
    Social Responsibility
Crang, M 39
Crash 69
creolization 15
cruise missile strikes, Afghanistan/Sudan
    122
cultural heritage trails (Vienna) 39

cybercities 45, 47
Cyberdemocracy 4
cybermobilities 44, 47, 53, 55, 56, 57;
  physical and virtual 11
cybernetics 131
cyberpublics 4
CyberSociety 4
cyberspace 4, 12

data: folding of 140–1; Internet Protocol
  (IP) 142
dataveillance 129
Dearnley, J. 4
de Certeau, Michel 29–30, 70
Delamaide, D. 127
De Landa, Manuel 56–7
Deleuze, G. 44
Department of Homeland Security (US)
  145–6
de Paula Gaheis, Franz 33
design paradigms 153, 171, 172; and text
  messages 166–8
design strategies 172
de-synchronization 7
differential mobility, surveillance 177–91
Digital Rights Management (DRM) 58n
directly observed therapy (DOT) 126
disaster potential 124
disaster recovery operations 121
disease transmission 2, 122–3
disruptions to journeys, bypassing 92–5
Dodge, M. 48, 52
Donald, J. 21
dot.com crash/dot.com boom 140
DOT *see* directly observed therapy
Douglas, M. 181
dread, concept of 124–5, 126
Ducatel, K. 86, 92
du Maurier, Daphne 79, 86
Durkheim, Emile 10

easyJet airline 46, 52
electronic travel information, 'real-time' 79,
  80, 81
Elle magazine 98n, 99n
Emerging Infection network (APEC) 128
enhancement, physical and virtual mobility
  108
Enlightenment 25
enterprise, cultures of 11
enunciation concept, and urban space
  representations 29

equivalence principle, civil inattention
  concept 172
Europe, tours through 24–5
extra-terrestrial intelligence, search for 128

fantasy, and film 12, 62, 64
Feather, J. 4
fiakers (hackney carriages) 27, 37
FIDS (flight information displays) 53
films, futuristic urbanism in 10
'fly-in' city 52
flyovers 62
Foucault, Michel 179
freeways 71
French, Shaun 46, 178, 181, 185
Fuller, G. 53

Galileo GPS system 130
gameplay 156
game-related proximity engagements
  158
geolocated technologies 13, 152, 153
get-at-ability, and accessibility 103,
  104–5
Gilpin, William 25
glasswerk.co.uk (website) 44, 48–51;
  mobilized events 50–1; virtualized
  performances 49–50
global cities 15
Global Emerging Infections Surveillance
  and Response System, Department of
  Defense (US) 128
globalization 122, 126, 127
global mobilities, and local risks 126–7
Global Positioning System (GPS) 69, 161;
  GPS-based geolocalization (2001–2002),
  Japan 161–2
Go, F. M. 123
Goddard, Jean-Luc 69
Godskesen, M. 83
Goffman, E. 170, 172
Gomes, Pete 148
Gosford Park 65, 72
Gothic Fractur script, street plates 32
Govers, R. 123
GPS *see* Global Positioning System
Graham, Stephen 2, 8, 14, 45, 47, 55–6,
  177–91
Grand Tours 24, 37; walking tours as part
  of 32
Great Britain Passenger Railway Timetable
  88

Greenwich Mean Time (GMT) 82
grid computing, SARS data 128
GSM cells 162
guidebooks (Vienna) 11, 23–4, 26, 29–31, 37–8, 40n; and walking tours 33, 38; *see also Vienna Baedeker*
Guillot, Romain 10, 13, 152–76

Hagerstrand, T. 104, 115n
Hartleben, guidebook of 31
Hatfield crash (2000) 85, 92
Haussman, Georges-Eugene 23, 38
Headrick, D. R. 129
Hepworth, M. 86, 92
Hertzian landscape 149; notion of 141, 142, 143, 144
Hetherington, K. 54
heuristic systems 188
hijacking, paradigm of 121
Holmes, David 47
Hong Kong: smart cards 130; Special Administrative Region, and SARS 123; transport systems 122
human-machine communication 91
hybrid collectives 179, 180, 183, 186
hybrid paradigms 154–5
hyper-mobility, and terrorism 122

ICTs *see* information and communication technologies
identity cards, electronic 128, 187
imaginaries 10
Immigration and Naturalisation Service Passenger Accelerated Service System (INSPASS) 184
Indices of Deprivation 103
information asymmetries 130
information and communication technologies (ICTs) 105–6, 152–76; and accessibility 104, 105; and Internet 104, 106; mobility effects 108; time-space opportunities 83–4
information flows, cities 1–2
informational mobility, and physical mobility 3
infrastructures 3, 11, 13; communication/telecommunications 46, 82, 129; military 129; mobility 85, 89; transport 8, 82, 90
inhibition techniques 188
InnovaCell (Mobilfone mobile phone operator) 160, 161, 162

INSPASS *see* Immigration and Naturalisation Service Passenger Accelerated Service System
instant messaging 173
Intelligent Transport Systems (ITS) 47
interactional mobility paradigms 155
'Interactive City' Conference (2004), Liverpool 51
Internet 149; adolescent prostitution, and Japan 173; base technology 131; and information and communication technologies (ICTs) 104, 106; Jindeo game *see* Jindeo (game); military origins 129; Nido game *see* Nido (Internet game); technologies 131
Internet Protocol (IP) data 142
Internet routers 14, 184
Internet users, software sorting 184
interviews, semi-structured 79–80, 86–94, 95, 96
Iraq: regime change in 129; western occupation 130; *see also* 9/11 attacks, 2001
ISPs (Internet Service Providers) 145

Jain, Juliet 12, 79–101
Jain, Sarah J. 10, 12, 61–76
Japan: adolescent prostitution (Internet) 173; GPS-based geolocalization (2001–2002) 161–2; KDDI (Japanese operator) 161, 162, 166, 169, 172; Tokyo-Yokohama conurbations 13, 169
Jencks, Charles 54
Jindeo (game) 155, 156, 157, 159; design team 159–61; development (2001–2003) 159–65; Japan, and GPS-based geolocalization (2001–2002) 161–2; MMORPGs and 163, 164, 165; Ubisoft trial and outcome (2002–2003) 162–5, 166, 167
John Lennon Airport (Liverpool) 44–5, 51–5; information, mobility 53–4; simulations 54–5
Johnston, C. 122
Joseph II 33
journey: as narrative 68; 'seamless', concept 80

KDDI (Japanese operator) 161, 162, 169, 172; Nido game 166, 169
Kenyon, Susan 12, 84, 102–20
kinetic elites 178, 184

Kitchin, R. 48, 52
Kizoom 80, 89–90, 91–2, 94, 95
Kling, R. 127
knowledge extraction, 1980s projects 129
'Konskriptionsnummern' (numbering of
    houses), Vienna 31
Koshar, R. 38
Kuhn, Thomas 153

Lamberton, D. 130
landscape, notion of 9
Langan, Celeste 71
Latour, B. 57, 178, 180
Laurier, E. 85, 96
Law, J. 9, 178, 180
Le Corbusier 22
Leibsoft 161, 162
Lianos, Michalis 178, 181, 182, 183, 187
Licoppe, Christian 10, 13, 152–76
Liebs, Chester 75n
Lindqvist, S. 122
linearity: European tours 24–5; linear cities
    21–2; linear movement, space defined by
    25–8; tracing of lines 22–5
liquid modernity, and mobility 3
Little, Stephen 13, 121–33
Liverpool 45–6; John Lennon Airport 44–5,
    51–5; Speke, first airport built in 45–6
Living with Cyberspace (J. Armitage and J.
    Roberts) 4
Lyons, G. 80, 88, 93

Mackenzie, Adrian 10, 13, 137–51
McKinley Conway, H. 52
McLay, G. 88, 93
Madonna 12, 61–7, 72
Manchester Airport 52–3
Manila, urban form of 62, 72
Marvin, S. 8
Massey, D. 56
massive multi-player online role-playing
    games (MMORPG) 154, 155, 156, 159,
    160; Jindeo 163, 164, 165
Master Hands 74n
Matrix, The 10
medical prophylaxis 123
Meitßl, G. 23, 36
Melbourne, City Link 47
metabolism, of cities 1
migrants, and cities 2
migration patterns 127
Miller, H. J. 105

Mitchell, Elvis 65
Mitchell, W. 46
Mitchell, W. J. T. 141, 142
MMORPG *see* massive multi-player online
    role-playing games
mobile office 96
mobile phones 84–5, 88–9; GPS locator
    function requirement (US) 130
mobile practices, emergent 91–7; blockages
    and disruption, bypassing 92–5; 'WAPing
    around' 95–7
mobile technologies 83
mobile workers 96
mobility: automatic sorting of mobilities
    183–5; business class 12; and cities 1–17,
    46–8; connected 44–8; differential
    177–91; free flowing 46; global
    mobilities, and local risks 126–7;
    informational 3; liquid modernity 3; low
    and high levels, and social exclusion 115;
    monitoring 127–9; paradigm 155, 171;
    permeable mobilities 183–4; physical *see*
    physical mobility; politics of 56; public
    character of 168–71; 'real-time' 79–101;
    reframing 122–4; and social exclusion
    12; socialized/interactional 172; spatial
    152; virtual *see* virtual mobility
mobility/moorings dialectic (Urry) 56
mobility turn, social sciences 152
modernization, and classifications 180
Mokhtarian, P. L. 107–8
Motion Picture Experts Group audio layer
    (MP3) 49, 57–8n
movement, images of 13
MP3 *see* Motion Picture Experts Group
    audio layer
multitasking 114
Muslim countries, terrorist attacks 130

National Commission on Terrorist Attacks
    128
National Intelligence Committee 126
National Rail Enquiries Service (NRES)
    96–7; call centre 88
National Rail Timetable 87
National Security Agency (NSA) 128
Naturalization and Immigration Services
    (US) 184
Neighbourhood Renewal Unit (NRU) 103
'network centric defense' system (US) 129
network society 15n; 1980s and 1990s 140
networking, personalized 7

networks: postal (Vienna) 36; schedules 7–8; Wi-Fi 145–6; wireless 140, 141–2, 149
new urbanism 1
Nido (Internet game) 154–5, 165; and equivalence principle 172–3; KDDI (Japanese operator) 166, 169; moral issues 173; on-screen encounters, and public character of mobility 168–71; screen *157*; switch to 155, 166–71; text messages, pivotal role in redefining design paradigms 166–8
9/11 attacks, 2001 13, 121; casualties 126; civil-military communication problems 129; consequences 122
nodes of connection 84
Non-places (M. Augé) 51–2, 70
Norris, C. 182–3
Novak, Marcos 47, 54
NRU *see* Neighbourhood Renewal Unit
NSA *see* National Security Agency
Nugent, Thomas 24

Ohmae, K. 126, 127
Ong, Mat 49, 50, 51
online activities, social implications 106–7
on-screen encounters 172; and public character of mobility 168–71
OpenPark Wi-Fi project (National Mall, Washington DC) 147–8
ordering: defined 180; and surveillance 181
organizations, and mobile technology 13
'Orientierungsnummern' (numbering of houses), Vienna 31
overflows 142–8, 149; collective-cultural-political 147–8; notion of 142; personal-infrastructural, in café and park 143–6; techno-practical 146–7
Owen, Clive 61, 62, 63, 64–5, 66, 68

Panini (card collection games for children) 166
paradigms: accumulation 167; civil 129; communication 154; design 153, 166–8, 171, 172; hijacking 121; hybrid 154–5; interactional mobility 155; mobility 155, 171; notion of 153
ParkBench TV project (London) 148
participation format-switching 157–8
passenger movement, simulation *55*
passengers, virtual 54
Patton, Jason 75n

Pearl Harbor, Japanese attack 121
Peel Holdings Ltd, and Liverpool Airport 52
Perrow, C. 124–5, 126
Perry, M. 84, 96
personalization/personalized networking 7, 8, 90
Peters, S. 85, 90
Pfund, N. 124
Philipp Haas & Söhne (Viennese department store) 39
physical mobility: and communication technologies 12; and informational mobility 3; and virtual mobility 12, 47, 102, 104, 106, 108, 109
Piccadilly Circus 144; Wi-Fi 'hotzone' 138, 139
pilgrimages, and touring 24
pre-planning 80, 84, 85, 88; meetings 95
preferred customers 8
Prelinger, Rick 74n
prioritization techniques 188
prison-industrial complex 7
Priss narrative, computer games 156–8
programmes of action, and social relations 179
'protestant work ethic' 81
proximity engagements, game-related 158
punctuality, notions of 85

Railtrack 88, 90
railways: 'railway time' 82; timetables 79, 82–3, 85–91; Victorian rail companies 86; Vienna 25, 26
Rajchman, John 48, 49
real and virtual, interplay between 10
'real-time' information and planning 79, 80, 81, 84, 92, 93, 94–5
Rebecca (D. du Maurier) 79
recombinant architecture 47
representational practices 11
research, mobility 3; further, need for 14–15
Ringstraße (Vienna) 23, 38, 39
Ringway Airport 52–3
risk, and complexity 124–6
Ritchie, Guy 12, 61, 65, 66
road pricing, electronic 183
Roberts, J. 4
roleplays, 'lifesize' 156
routers, Internet 14, 184
Ryanair airline 46, 52, 53

197

St Stephen's Cathedral (Vienna) 30, 39
Salomon, I. 107–8
Salvador, T. 84
SARS *see* Severe Acute Respiratory System outbreak, East Asia (2003)
'scapes' 9
scheduling 7, 12, 81, 85; 'do-it-yourself' 95; fixed, concept of 89
Schipol airport, Amsterdam 184
Schivelbusch, W. 25
Schmidl, A. 33, 37
Science and Technology Studies (STS) 46
'seamless journey', concept 80
'seaport, to e-port' movement 51
security, military definition 129
September 11 2001 attacks *see* 9/11 attacks, 2001
Serres, Michel 57, 179
service set identifier (SSID) 144–5
SEU *see* Social Exclusion Unit
Severe Acute Respiratory Syndrome (SARS) outbreak, East Asia (2003) 13, 121, 122–3; global response to/global tracking 128, 130; grid computing 128
Sheller, Mimi 1–17
Sherry, J. 84
sightseeing tours *see* tourism and travel
Simmell, G. 152
Simons, Derek 70
simulations 12, 54–5
Sitte, Camillo 26
Sloterdijk, P. 184
smart cards 187; Hong Kong 130
SMS messaging 53
social control: city 181–3; new 178
social exclusion 102–20; and accessibility, lack of 103–4; in cities 103; concept 102; definition 103; mobility-related 12, 115; origins 102; and virtual mobility 106–7, 114
Social Exclusion Unit (SEU) 103
social network analysis 129
social relations, and programmes of action 179
Social Worlds Theory 178
Society for the Protection and Preservation of Artistic Monuments 23
socio-technologies 3
software, self-programming 188
software-sorting 184; systems, politics of 185–7

Soria y Mata, Arturo 22
spaces: civic 36; as non-places 51–2; public 35, 36; tourist 35; urban *see* urban space
spatial mobility 152
spatio-temporal order, manipulation of space in 69
Speke, first Liverpool airport built in 45–6
Splintering Urbanism (S. Graham and S. Marvin) 8
Sports Utility Vehicles (SUVs) 64, 71, 75n
Spring, Ulrike 10, 11, 21–43, 82
SSID *see* service set identifier
Stadtbahn (city train), Vienna 23
'stake-holder' organizations, British 79
Star, S. 178, 180, 187
Star (six-minute film) 12, 61–74; and automobility 68, 69, 71; Blur 2 song 69; BMW depicted in 61–2, 64, 65, 66, 67–8, 72; description 63–6
Stoler, Ann 73
substitution, physical and virtual mobility 108
substitution hypothesis, mobilized events 50
Suchman, L. 91
Sudan, cruise missile attacks 122
surveillance: actor-network theory 14; automated systems 182, 185; conventional systems 186; differential mobility 177–91; and ordering 181; permissive mobility 130; price of mobility 122
surveillance society 129
Switzerland, walking tours 32
synchronization: de-synchronization 7; time-space 81
systems theory 131

Tadiar, Neferti 62, 72
techno-economic networks (TENs) 180
technologies: airports 14; and co-presence 122; geolocated 13, 152, 153; Internet 131; and landscapes 9; mobile 13, 83; sociology of technology 14; transport 11; *see also* socio-technologies
technoscapes 8–9, 10
technospaces 11
teleworking 126; mobility effects 109
TENs *see* techno-economic networks
terrorism: air travel 13; bio-terrorism 129; definitions 129; fears of 2; and hyper-mobility 122; Muslim countries 130; responses 129–30

text messaging 172; and design paradigms 166–8
3-D simulations 54–5
Thrift, N. 10, 11, 46, 178, 181, 185
time: clock 81, 85; compression of 37–8; Greenwich Mean Time (GMT) 82; wasted 92; work 82; *see also* 'real-time' information and planning
'time = money' concept 81, 85
timetables, railway 79, 82–3, 85–91
TOC (Train Operating Company) 87
Tomen Telecom (KDDI sub-contractor) 161, 162
topology 186, 187
tourism: and business travel 2; concept of tourist 35, 40n; Grand Tours *see* Grand Tours; mass tourism 11; and pilgrimages 24; spaces, tourist 35; in Vienna 11, 22–4, 28, 32–5, 38, 39; walking tours 32–5, 38, 39
tramps 2
transarchitectures 47
translation regimes, and co-ordination regimes 180
transmateriality 51, 57n
transmigrants 2
transportation systems 1–2, 5; car, as mobile commodity 10, 62, 63; in Hong Kong 122; imperial origins 129; Intelligent Transport Systems (ITS) 47; railways *see* railways; studies 3; in Vienna *see* transport (Vienna)
'Transport Direct' (government-sponsored organization) 80, 86, 93–4
transport (Vienna): City Train *27*; driving systems 26–7; fiakers (hackney carriages) 27, 37; public 23, 36; railways 25, 26; tramway system 27; urban traffic 26
travel: business 2; real-time information 79, 80, 81, 92, 95; *see also* railways; tourism; transportation systems; transport (Vienna)
tuberculosis (TB) 125, 126
Turner, J. M. W. 25

Ubisoft trial, Jindeo game (2002–2003) 162–5, 166, 167
United States: Air Traffic control system 129; cellphones, GPS locator function requirement 130; Center for Disease Control, Atlanta 128; Department of Homeland Security 145–6; Global Emerging Infections Surveillance and Response System 128; National Commission on Terrorist Attacks 128; National Security Agency 128; Naturalization and Immigration Services 184; 'network centric defense' system 129; OpenPark Wi-Fi project (National Mall, Washington DC) 147–8; Project for an American Century 129; wireless communication systems, third-generation 8
urban culture 14–15
urban form 10; and violence 12
urban jazz 84
urban mobility systems 5–8; historical aspects 6
urban new media studies 10, 13
urban planning 38
urban space: American 62; framing 11; interaction with (designers and game-users) 13; and mobile communication 4; tourism 35; violence of 61
Urry, John 1–17, 56

von Geusau, Anton Reichsritter 30–1
Vienna (nineteenth-century) 11, 21–43; areas, main 28; bird's eye view 29; extension (1850s onwards) 22–3; growth 35–6; guidebooks *see* guidebooks (Vienna); house numbering 31–2; linearity/linear movement 21–8; linking of city 38–9; as metropolis 23; parks, opening to public 33; planning competition (1858) 23; public space, moving across 35–7; relational perspective 26; 'representation crisis' 28–32; Ringstraße 23, 38, 39; sightseeing tours 11, 22–4, 28, 32–5, 38, 39; streets and squares 26, 28, 29, 32; time, compression of 37–8; transportation system *see* transport (Vienna); *Vienna Baedeker* 23–4, 32, 35, 37; walking in 32–5
violence 12, 61, 73
Virilio, P. 6, 84
virtual communities 47
virtual mobility: accessibility enhancement 113; and physical mobility 12, 47, 102, 104, 106, 108, 109; and social exclusion 106–7, 114; as substitution for trips 109
virtual movement 56
virtual passengers 54
virtual reality 12

Wagner, Otto 26, 28
Wagner, Richard 64, 69, 74n
Wajcman, J. 108
walking tours 32–5, 38, 39; excursive
  walking 33
Wanderungen (tours on foot) 32
WAP (Wireless Application Protocol) 161
WAPing 81, 91, 92, 95–7; concept 97;
  implications for public transport
  providers 97–8
war on terror 122, 128, 129–30
war-chalking, practice of 145, 150n
'wasted time' 92
Webber, M. 127
Weber, T. 138
Weidmann, F. C. 33
Welford, R. J. 126
Wellman, Barry 4–5, 7, 84
WHO *see* World Health Organisation
Wi-Fi 13, 137–51; networks 145–6;
  OpenPark project (National Mall,
  Washington DC) 147–8; Piccadilly
  Circus, 'hotzone' 138, 139

Winseck, Dwayne 184
Wired World (J. Dearnley and J. Feather) 4
wireless communication systems, third-
  generation 8
wireless internet service providers (WISPS)
  146
wireless networks 140, 141–2, 149
Wireless Venture Association (WIVA)
  161
WISPS *see* wireless internet service
  providers
WIVA *see* Wireless Venture Association
Wolfowitz, P. 137
Woods, David Murakami 14, 177–91
work time, demands 82
World Commission on Environment and
  Development (1987), Bruntland Report
  126
World Health Organisation (WHO) 128
World Trade Center, attacks on *see* 9/11
  attacks, 2001

'zebra strategies' 127

For Product Safety Concerns and Information please contact our EU
representative GPSR@taylorandfrancis.com
Taylor & Francis Verlag GmbH, Kaufingerstraße 24, 80331 München, Germany